THE OLI ...
THE BETTER WE WERE

MSgt A. A. Bufalo USMC (Ret)

Author of

SWIFT, SILENT AND SURROUNDED

ISBN 978-0-9745793-1-3

First Printing – February, 2004
Printed in the United States of America

www.AllAmericanBooks.com

IN MEMORY OF

First Sergeant Ed Smith and all of the warriors who have died in the service of our country while fighting the war on terror in Afghanistan, Iraq and elsewhere.

PREFACE

The response to my first book, *Swift, Silent and Surrounded,* was overwhelming. I received letters, phone calls and e-mails from Marines and former Marines all over the world telling me their thoughts about a favorite story, or on the book as a whole. Almost without exception the question was asked, "When is the next book coming out?" Now keep in mind, I had never really considered the possibility of writing another volume, but eventually decided, "Why not?" And here it is!

But all of the feedback wasn't positive. A friend of mine who happens to be a very proud former WM wrote a long and impassioned e-mail in which she expressed displeasure with my opinions concerning women in the Marine Corps. I am happy to report, however, that she went back and reread the chapter in question and came to realize I was not questioning anyone's dedication or patriotism. I was simply questioning the wisdom of policies which either provide for a double standard, or lower standards altogether, in order to accommodate women. I knew I was stirring up a bit of a hornet's nest, but as we all know nothing ever gets accomplished by maintaining the status quo.

One small disappointment connected with *SS&S* concerned Leatherneck Magazine's refusal to do a book review. I had been approached by a former Marine named Red Bob Loring, who has done many fine reviews for that publication, about doing one of my book. In order to facilitate his request a copy of *SS&S* was sent to the Editor of Leatherneck, and I found his response to be somewhat troubling. He informed Mr. Loring his review idea had been declined, not because the writing was bad or the cover was ugly, but because the content "did not support Marine Corps policies." I am not thin-skinned, and could have easily accepted rejection based upon my lack of skill as a writer since I harbor no illusions about my ability. But to say it is necessary to "toe the line" in order to receive a review smacks of Pravda. I thought to myself they may as well limit reviews to FMFMs in the future, since those contain nothing but USMC doctrine. Something tells be *this* book may not make the review list either.

The title, "The Older We Get, the Better We Were," refers to a phenomenon many of us experience as we get older which causes us to remember ourselves as being taller, stronger and better looking in our youth than we really were. It also points out how we tend to remember the good times from our years in the Corps, and how our memories of the negative things tend to fade more and more with each passing year. I can say from personal experience these occurrences are all too real.

This volume contains a few more political stories than the first, but all of them relate in some way to the Marine Corps, leadership, or military policy. Many concern the Presidency. I decided that was appropriate since the President is also the Commanded-in-Chief, and as such sets the tone for the armed forces as a whole and it just made sense to underline the traits we Marines admire in a leader, as well as the things we loathe.

The format of this book is a little bit different from the first. This time I have included profiles of a few well known Marines, and have included a few articles written by or about non-Marines which I believe we all can relate to in some way.

I have used a variety of sources to compile the work that follows, and it is quite possible the reader may come across a few pieces which have been circulated on the internet or by word of mouth. That is intentional. I have discovered many "nuggets" out there which I believe should be kept alive, and many of them have been catalogued in my two books with an eye toward preserving them for posterity.

I hope you enjoy what follows, and that in some small way it helps make our Brotherhood just a bit stronger. Semper Fi!

Riverview, Florida
October 20th, 2003

TABLE OF CONTENTS

BRANDED

Kristine Kirby Webster

"We are what we repeatedly do. Excellence, therefore, is not an act, but a habit." - Aristotle

This article written by Kristine Kirby Webster was read by Brigadier General William D. Catto at the 227th Marine Corps Birthday celebration at the Naval Postgraduate School on November 1, 2002.

The other night I was sitting and knitting, and working mentally on a presentation I am putting together. Specifically, I was trying to encapsulate in a few points the hallmarks of a great brand. I decided that a great brand is enduring, establishes affinity, and engenders loyalty.

After mulling over these hallmarks, I found myself wondering what I would consider to be the Great American Brand. Would it be Sears, the original catalog powerhouse? All the Ma Bells, the forerunners of telecoms today? Would it be McDonald's and their ubiquitous arches? How about Coke and their national and global reach?

No. The great American brand, in my eyes, is the United States Marine Corps.

Now, I can almost hear many of you saying, "Wait just a minute, Kristine.... the Marines don't sell anything! How can it be the Great American Brand?"

I admit it. When most people think of branding, they think of it as a part of a sales plan, one designed to generate profits. But brands needn't be about sales. As the hallmarks of a great brand demonstrate, the bond and the relationship created is the most important goal of a brand. It can't be stated enough: the true promise of a brand is only realized through the customer-brand experience and the resultant relationship.

The Marines are the smallest of the U.S. military services. But if you were to gauge size merely by the number of bumper stickers on cars across America, the Marines would win hands-down as the *largest*.

And the Marines aren't content to simply rest on their historic laurels. They consistently promote their brand through multi-channel marketing efforts (both externally and internally) more precisely and effectively than any other service, and many organizations.

Perhaps that is why they have numbers such as the following that would make any for-profit business jump for joy:

The Marines have consistently met their monthly recruiting goals for more than seven years running.

For the Fiscal Year 2003 (which started October 1, 2002), there are 6,100 openings for Marines wishing to re-enlist during this year. As of October 11, 2002, more than 5,100 Marines had requested re-enlistment. At that rate, three weeks into their fiscal year they would meet their annual goal. (Talk about excellent retention!) One of the main functions for success in branding is consistency. The Marines have had some form of the Eagle, Globe, and Anchor logo since their founding in 1775, and the Commandant of the Marine Corps always has the license plates "1775" on his vehicle.

Almost all Marines begin or end all conversations, correspondence, etc. with "Semper Fi," their motto ("Semper Fidelis," meaning "always faithful").

And, of course, who can forget the Marine Corps bulldog? All of these symbols combine to reinforce the brand and serve as markers of loyalty and a sense of community.

In their book *The 22 Immutable Laws of Branding*, Al and Laura Ries note that "if the entire company is the marketing department, then the entire company is the branding department."

This is absolutely true of the Corps. Each Marine is a walking, talking advertisement, and a persuasive one at that. The Marines understand the importance of their brand - both externally and internally - more than any other service, and more than most companies. To the Marines, their brand is living, breathing, historically-based and constantly evolving.

Every strong brand today recognizes that the brand is not a static thing; it needs to be constantly evolving to meet the needs of their customers, and it needs to be nurtured and promoted in order to endure.

The Marines understand the need to go out and find those Marines of tomorrow. They are sponsors of such events as the X Games, NASCAR, NFL Football, and other sporting events that are attractive to their target audience.

They don't just sit around waiting for candidates; they use the proactive nature of their brand and message and mission to go out and attract people who want to be Marines. They promote not only the tangible benefits of the brand - the uniform, the respect, the ability to serve your nation, and a chance to see the world, but also the intangible - the feeling of pride, of belonging to a select group, of aspiring to be someone great. (Another Marine tag line is "The Few, The Proud, The Marines.")

They also use their proactive nature to "keep" the Marines who have served in the past. Have you ever heard the oft-said phrase, "Once A Marine, Always A Marine?" The

Marines make great efforts to retain the affinity and relationship between the Corps and the Marine even after a person's active service is over.

To this end, they have a program called "Marine for Life." The Marine for Life program's mission "is to provide sponsorship for our more than 27,000 Marines each year who honorably leave active service and return to civilian life in order to nurture and sustain the positive, mutually beneficial relationships inherent in our ethos 'Once A Marine, Always A Marine.'"

The Marines clearly understand the importance of relationships, longevity, and of loyalty. Besides being a force to be reckoned with on the battlefield (pun intended!), they are a force to be reckoned with off it. They have a large contingent of Marines - past and present - as well as their families, whom they can rely on to promote the needs and the vision of the Corps, from the halls of government to the smallest farm communities, from inner cities to Fortune 500 boardrooms. The amazing reach of their message is only superseded by their consistency of purpose and message.

What can your brand learn from the Marines?

That consistency is vital, that loyalty is a valuable asset, and that relationships created in the brand promise, and delivered on by the brand fulfillment, are lasting. That treated well, you can create and have customers for life.

So, perhaps your brand needs to go to boot camp and learn some brand promotion and loyalty techniques from the Marines.

Is your brand up to the challenge?

Kristine Kirby Webster is President of The Canterbury Group, Ltd. a full service direct marketing agency and consultancy which specializes in non-profit fundraising and relationship marketing programs.

DEADLY DAY
For Charlie Co

Rich Connell and Robert J. Lopez

"Walk softly and carry an armored tank battalion."
– Fictional Colonel Nathan Jessup from the film "A Few Good Men"

During Operation Iraqi Freedom the Marines got the tough slogging on the way to Baghdad, much the same way they got the tough assignments from MacArthur during the Korean War - and that's the way it should be. Army generals may be jealous, and even resentful, of the Marine Corps, but they rarely fail to use us to good advantage when given the opportunity.

This story is a good example of the fog of war and demonstrates how quickly decisions must be made in combat, and how easy it is to become involved in a "friendly fire" incident. That is especially true when the support is coming from non-Marine units.

One of the big revelations is the connection between the battle fought by Charlie company and the now infamous Jessica Lynch rescue. If Marine tanks had been available for this fight as originally intended there is a good chance many of the casualties suffered by Charlie Company could have been avoided. But while the press was focused on the rescue of one female POW, and later on what she was eating in the hospital, Marines were dying anonymously as a result.

I feel a bit of a personal connection to what happened to Charlie Company for a couple of reasons. My first tour of duty was with AAVs - back then we called them Amtracs - and I always said they were nothing but big targets. I hated riding around inside of those deathtraps, and was very thankful when my transfer to Recon came through. That duty was of course dangerous in a different way, but at least I could see it coming.

The second thing is the involvement of the A-10 Thunderbolts. My Godfather worked for the company that built those planes, and I have visited the assembly line on a number of occasions. I am hopeful that incidents such as the one contained in this article never happen again:

The convoy rumbled north, through the heart of the Iraqi city of Nasiriyah. It was the fourth day of the war, and the men of Charlie Company had orders to capture the Saddam Canal Bridge on the city's northern edge.

The Marines were taking heavy fire. Then there was an ear-splitting blast. A rocket-propelled grenade ripped open one of the amphibious assault vehicles, lifting it off the ground. A thick, dark cloud filled the vehicle's interior. Some of the Marines donned gas masks, fearing a chemical attack. Screams pierced the smoke.

"We got a man down! We got a man down!"

The Marines' light armor had been pierced, and with it any illusion that this would be easy. They would take the bridge, but at a cost. Eighteen men from a single company were killed that day and fifteen wounded, making it the deadliest battle of the war for U.S. forces.

Public attention, briefly riveted on the fighting in Nasiriyah, has since moved elsewhere. The struggle to rebuild Iraq and contain mounting guerrilla violence now occupies center stage. But the Marines of Charlie Company, now back home, are not ready to put that Sunday in March behind them.

They want to know why commanders sent them into an urban firefight without tanks, without protective plating for their vehicles and with only half the troops planned for the mission.

They want to know why an Air Force fighter strafed their positions as they struggled to hold the bridge, killing at least one Marine and possibly as many as six.

Five months later, the U.S. Central Command was still investigating the "friendly fire" episode. The Marine Corps conducted its own review of the battle but said it would not release its findings until the other investigation were finished.

The *Times* reconstructed the battle from interviews with eleven Marines who fought that day. Their accounts paint a gory and chaotic picture of ground combat that contrasts with the many images of U.S. forces using precision bombs and long-distance weaponry against an enemy that quickly abandoned the fight.

In Nasiriyah, Iraqis stood their ground and threw all they could muster at the leading edge of the American forces. By day's end, the price of controlling the road to Baghdad had become gruesomely clear to both sides.

Charlie Company had reached Nasiriyah after pushing up eighty-five miles from Kuwait. Another Marine unit had seized a bridge leading into the city over the Euphrates River.

Charlie Company's mission on March 23 was to take a second bridge three miles north. Controlling both spans was crucial to moving a massive Marine Expeditionary Force to Baghdad. Had things gone as planned, the two hundred Marines in their lightly armored vehicles would have avoided the densely populated heart of Nasiriyah, a city of 500,000. They were supposed to take a roundabout route to the north bridge, swinging east of the city behind a dozen M1-A1 Abrams tanks and a second Marine unit, Bravo

Company. But Bravo Company's vehicles sank several feet deep in mud flats east of Nasiriyah. Its two hundred men could not help take the bridge.

The tanks were also out of the fight, diverted on a rescue mission. The Army's 507th Maintenance Company had taken a wrong turn that morning and been ambushed near the city. Eleven soldiers were killed and seven captured, including PFC Jessica Lynch.

The Marines' tanks rushed to retrieve survivors, burning their fuel in the process. When they returned, they were sent to the rear to refuel just as Charlie Company was preparing to push north.

"Where the hell are the tanks going?" Corporal Randy Glass recalled thinking. "Why the hell aren't the tanks in front of us?"

Despite the lack of armor and the stranding of Bravo's men, Charlie Company was ordered to take the north bridge and to get there by the most direct route - a three-mile stretch of highway lined by buildings and alleyways. Some intelligence reports called it "Ambush Alley."

Lieutenant Colonel Rick Grabowski, the battalion commander, said that going ahead made sense at the time. Though concerned about Ambush Alley, commanders did not anticipate a tough fight for the bridge, he said: "None of us really knew what was on the northern side of the city."

And time was of the essence. If they waited for the tanks to return or for troop reinforcements, the Marines risked fighting for the bridge in darkness, Grabowski said.

There was another factor driving the Marines forward that day. It reflected a state of mind as much as the state of the battlefield. "Keep moving" was the motto of Charlie Company's battle regiment.

"Once we were in the city and we made contact," Grabowski said, "there wasn't going to be any backing down."

When he got the order to move into the city, Sergeant William Schaefer thought he'd heard wrong.

"Say again," he called into his radio.

Schaefer was a commander at the head of Charlie Company's eleven amphibious assault vehicles. The men inside were from the beach towns of Southern California, the hamlets of upstate New York and many places in between. One planned to enroll at Rutgers University in New Jersey when he got home. Another wanted to be a Reno cop. Some were immigrants. Others were from proud military families.

They were part of the 1st Battalion, 2nd Marine Regiment and had shipped out in January from their base at Camp Lejeune, N.C. Their tub-shaped assault vehicles, called "tracks," are a thirty-year-old design made for taking and holding beachheads. They are twenty-six feet long, carry up to twenty Marines each, and are armed with .50-caliber machine guns and grenade launchers. Their reinforced-aluminum skin is vulnerable to artillery and rocket-propelled grenades, or RPGs - unlike the heavy armor on tanks. Thick steel plating can be attached to the tracks, but none was available to outfit Charlie Company's vehicles when they reached the war zone.

"Eight Ball, Oscar Mike," Schaefer barked into the radio, and with that signal the company was on the move.

The tracks crossed the Euphrates on the bridge captured earlier and moved single file up Ambush Alley. It was a little before noon. On both sides, a dense warren of mud-colored buildings pressed up against the road. At first, the Iraqis seemed to welcome the Marines. A few waved white flags. Then, in a breath, the convoy was under attack from all directions. Iraqis were firing from rooftops, from around corners, from machine-gun nests hidden in side streets.

"We saw women shoot at us with RPGs. We saw children shoot at us," recalled the company commander, Captain Daniel J. Wittnam. "We never saw one person in uniform."

Returning the fire, Schaefer alternated between his machine gun and grenade launcher, working a foot pedal that spun his turret right, then left. Schaefer, of Columbia, South Carolina, said the Marines tried to distinguish between Iraqi fighters and noncombatants. "But at that point, it was hard." The enemy, the Marines learned later, was a combination of Iraqi army soldiers, Fedayeen Saddam militiamen and Baath Party loyalists.

One man knelt and aimed an RPG at Schaefer's track. A burst of .50-caliber fire cut off the top half of the Iraqi's body.

"Pieces of people were all over the street," said Lance Corporal Edward Castleberry, who was at the wheel of Schaefer's vehicle.

Near the rear of the convoy, Sergeant Michael E. Bitz was driving a track crowded with more than twenty Marines. Bitz and his crew had picked up extra men when the company's twelfth track broke down outside town. Men were crammed on bench seats amid boxes of ammunition. Several were riding atop the vehicle.

In the middle of the column, Marines on another track shouted for more firepower to answer the torrent of incoming rounds. Lance Corporal Eric Killeen, a weightlifter from Florida's Gulf Coast, popped out of the hatch with his fifteen-pound squad automatic weapon, a machine gun that can spray a thousand rounds per minute. Killeen poured fire down side streets, into doorways, at second-story windows.

"My adrenaline was pumping so high," Killeen said. "Every emotion you can imagine was running through my body."

Castleberry, a Seattle snowboarder who'd joined the Marines the day after the September 11 terrorist attacks, steered the lead vehicle with one hand and fired his M-16 rifle with the other.

"I figured one more gun couldn't hurt," he said.

The convoy pushed north, the tracks pausing and pivoting at times to allow gunners a better view. They were almost through the gantlet of Ambush Alley. Their objective, the Saddam Canal Bridge, was a few hundred yards away.

Inside Bitz's overcrowded track, it was dark and noisy. The air reeked of diesel fumes. Marines were on top of one another. Some stood on the shoulders of their comrades, firing M-16s from a hatch near the rear.

Glass, a twenty-year-old from Pennsylvania who had joined the Marines hoping to see combat, was sharing a menthol cigarette with Sergeant Jose Torres when an explosion lifted the twenty-eight ton vehicle into the air. "Immediately, I went deaf," Glass recalled. An RPG had punctured the track's aluminum body - and with it the Marines' faith in their technological edge. Their tubs were not meant for this kind of fight - especially without the bolt-on armor plating.

"My eyes! My eyes!" shouted Torres, temporarily blinded.

"Glass is dead!" someone screamed in the chaos.

Glass wasn't dead, but his left leg was a bleeding mass.

Up top, the explosion had torched rucksacks tied to the track, turning them into balls of fire. Bitz drove the burning vehicle forward. This was no place to stop.

"Keep it tight! Keep it tight!" Schaefer shouted into the radio, not wanting any stragglers left behind.

The tracks finally crossed the Saddam Canal Bridge, a nondescript concrete span over an irrigation channel. Though it seemed an eternity, the trip had taken only a few minutes. On each side were swampy irrigation ditches and brush. Beyond was open flatland. The road was raised, and Charlie Company was an easy target.

The tracks fanned out over a quarter-mile-wide area. Marines in charcoal-lined chemical suits and Kevlar flak vests poured out of the vehicles and sought cover on both sides of the road. They had taken the bridge. Holding it was another matter. Small-arms fire exploded from the fields to the east and west and from the city to the rear. Marines scrambled out of Bitz's burning track. A release on the rear loading ramp didn't work, so the men piled out through a small hatch, climbing over the wounded.

Schaefer helped carry out Glass, whose left leg had been tied with a tourniquet. Bitz was carrying injured Marines to cover when a shell exploded, spraying him with shrapnel. Blood streamed from his face and back as he continued hauling the wounded to safety.

"He was acting like nothing was wrong," Schaefer said.

A plume of black smoke rose from Bitz's track. Mortar shells landed on each side. The Iraqis quickly adjusted their aim and slammed the vehicle. Burning ammunition began punching out the track's sides.

Corporal William Bachmann, a New Jersey skateboarder, was wedging his lanky frame into a nearby depression when he saw a flash of light from the vehicle. A large-caliber round flew past him. "If I was standing up," he said, "I would have been hit."

Corporal Randal Rosacker of San Diego set up his machine gun, providing cover for other Marines. His is a military family - his father is chief of the boat on a Navy submarine. Rosacker was cut down by an Iraqi artillery round or mortar shell, and was one of the first Marines to die that day.

The company's fifteen-man mortar squad set up a row of tubes on the east side of the road. The squad had no time to dig fighting holes. The Marines worked their three tubes furiously, knocking out Iraqi mortar positions across the canal. They fired so many rounds so quickly that their mortar tubes were glowing, almost translucent.

Outnumbered by the Iraqis' mortar positions, the squad was a prime target. Incoming shells were landing closer and closer. Finally, the Iraqi mortars found their mark. Nine members of the squad would die before the battle was over.

Second Lieutenant Frederick Pokorney Jr., a 6-foot-7-inch former basketball player, tried to call in artillery strikes on the Iraqis. He was the company's forward artillery observer, and when he had trouble getting through on his field radio he moved to higher ground for better reception. An RPG hit him in the chest, fatally wounding him.

As the casualties mounted, Wittnam wanted helicopters to evacuate the wounded. But there was "No way in hell..." they could land. "It was too hot."

Navy Corpsman Luis Fonseca was giving morphine to Glass and another wounded Marine in one of the tracks. With a black marker, he scrawled "1327" on Glass' head, indicating the time the painkiller was administered. The medic ran up and down the road looking for wounded when he saw Wittnam.

"We're starting to win this battle," Fonseca recalled the captain saying. Fonseca wasn't convinced. "I know there's a bullet with my name on it," he recalled thinking. "I'm gonna do my job until I get hit."

Machine gunners needed more ammunition. Sergeant Brendon Reiss, a squad leader, and Corporal Kemaphoom Chanawongse, a Thai immigrant from Connecticut, ran to get more ammo boxes from one of the vehicles. An artillery round exploded, killing Chanawongse and fatally wounding Reiss.

Around 1330 Schaefer decided to evacuate the wounded, even though it meant going back through Ambush Alley. All eleven tracks had made it across the bridge. Schaefer lined up six of them in a column to head south.

"I was willing to take a chance because we had guys bleeding to death," he said. "I was tired of seeing people getting killed."

Bachmann and Lance Corporal Donald Cline, a former surfer raised in La Crescenta, were firing from behind a mound of earth. Word came that volunteers were needed to load the wounded onto the vehicles.

"I'm going to help" Cline said, running toward the tracks spread out north of the bridge. It was the last time Bachmann saw his friend alive.

The Marines heard the plane before they saw it. The Air Force A-10 Thunderbolt, known as the "Warthog," flies ground-support missions, using its heavy gun. Corporal Jared Martin, a former high school wrestler from Phoenix, was outside Schaefer's track when he heard the growl of the jet fighter's twin engines. Its 30-millimeter cannon, which can shoot 3,900 rounds per minute, whipped up dense clouds of sand.

"He was low," Martin said. "He was coming right toward us. The next thing I know I'm feeling a lot of heat in my back." Blood streamed from his right knee and left hand. A piece of shrapnel lodged below his eye. "My fingers, they were pretty much dangling," Martin said.

Lance Corporal David Fribley of Florida was just steps from the cover of Schaefer's vehicle when rounds from the A-10 tore into his chest. "I wore what was inside of his body on my gear for a couple of days," Martin said.

To ward off the friendly fire the Marines shot flares, which streaked the sky with green smoke, but the A-10 made several strafing runs before it broke off the attack.

Schaefer hoisted a U.S. flag on his turret. He hoped the Warthog pilot would see it and hold his fire. He also wanted the tracks behind to be able to keep him in sight. "Watch for the flag," he radioed to the convoy of six vehicles heading south with the wounded.

As the column started back toward Ambush Alley, one of the AAVs exploded. Inside another track, Marines heard bullets bouncing off the aluminum skin. Glass, who had already been in one track that broke down, turned to Corporal Mike Meade, whose leg was also injured.

"If this track stalls we're getting out," Glass said. "It's a death trap." A minute later, the vehicle stopped. Glass and Meade struggled out.

Fonseca, the medic, heard the whistle of incoming shells and shoved a sergeant on top of Glass and another injured Marine. Then he piled on top to give added protection. Three RPGs flew by and exploded about a hundred feet away.

"I need to save these boys," he recalled thinking. "I need to take them back home."

Glass saw an A-10 fire on one of the tracks. It's unclear whether it was the same jet that had flown over earlier, since two of the aircraft appeared to be operating in the area.

"The A-10 came down hard and lit the track up," Glass said. "There's no mistake about it."

Torres was lying nearby when he saw the jet bearing down on him. "It was slow motion," he said. "I turned at the last moment to avoid a direct hit."

Still, the Warthog's rounds tore through his left side. "When he pulled the trigger," Martin said of the pilot, "it was just a wall of blood." Grabowski, the battalion commander, said that as many as six Marines may have been killed by A-10 fire. Wittnam believed it was one.

Schaefer's convoy, now down to five vehicles, was crossing the bridge. In front, a track that normally carried the mortar squad had several Marines inside. As the track came off the bridge, an Iraqi shell dropped down the left-side cargo hatch, ripping the vehicle in half.

"A hand and arm bounced across the front of my vehicle," said Schaefer. The remaining vehicles raced around the burning track. The rear of one track was crushed by an Iraqi shell, killing the wounded Bitz. The driver kept going. The four surviving tracks made it to Ambush Alley. They were met by gunfire from all sides.

Bullets ripped through Schaefer's transmission fluid tank, and Castleberry felt the steering wheel freeze. "Hold on!" he shouted over the intercom as the track careened toward a light pole. Castleberry gunned the 525-horsepower diesel engine, hoping to knock down the pole, but the track slammed to a halt and swung to the left toward a two-story concrete house.

An RPG blew away the track's front hatch, six inches above Castleberry's head. Stunned, his face and hair singed, he jumped into the street. Schaefer radioed to the three surviving tracks: "Don't stop. Keep going."

Inside the disabled track, a dozen Marines grabbed ammunition containers and the wounded and headed for the house.

Schaefer and two other Marines, one injured, were pinned down outside the track. “Then all hell broke out,” Schaeffer said. “They just started coming out of nowhere, hundreds of them.” Iraqis were charging the Marines. Schaefer aimed his M-16 and quickly used up two thirty-round magazines as he killed some of the attackers and forced others to take cover.

“When you’re scared,” he said, “you pull your finger pretty fast.”

Two of the other Marines, meanwhile, scaled an eight-foot wall and went into the house. An Iraqi man and woman ran out the back door. The Marines hoisted the wounded over the wall and put them inside. The windows of the house were hidden by piles of sandbags and sacks of flour. In one room were pictures of Saddam Hussein and a man who looked like Jesus.

Out on the street, Schaefer and the two other Marines were holding off the advancing Iraqis. Then the driver and a crewman from the track that had been ripped in half at the bridge appeared in the street. One was blinded. The other was limping.

“We’re laying cover fire for them,” Schaefer said, “and they hobbled inside.” Schaefer was on his last magazine. ‘This is it,’ he recalled thinking. ‘They’re going to overrun me. Then he heard the roar of a track driven by Corporal Michael Brown. He had disregarded Schaefer’s instructions to continue and had turned around. Scooping up the three Marines, Brown took off in a rain of enemy fire.

“He saved my life,” Schaefer said.

About seven Marines took up positions on the roof of the house. Martin, his wounds patched up, spotted two men peeking around a corner with an RPG. He fired and they fell. Martin looked at his watch. It was about 1500. “We have about two hours before the sun goes down," he thought. "Then we're gonna be really screwed."

The man who lived in the house burst through the back door yelling. He entered the room where the wounded were being guarded, Castleberry said, and was shot dead.

A lance corporal with the only operable radio called other units at the south end of the city for help, but the battery was low and he couldn't tell whether the message was getting through. With Iraqis now twenty yards from the building, the Marines on the roof were going through hundreds of rounds. Castleberry had fired so many grenades from his M-16 that the plastic hand grip on his launcher was melting.

Two Iraqis sped by on a motorcycle, the passenger firing an AK-47. On a second pass, one Marine hit the driver, spilling the bike. As the gunman tried to escape on the motorcycle, Castleberry unleashed another grenade. He saw a flash and the man's body blew apart.

Ammunition was running low. Castleberry and another Marine dashed to the disabled track, grabbing antitank missiles and crates of bullets as they dodged enemy rounds. On the roof, Martin and other Marines were trying to use shards of broken glass to reflect sunlight and get the attention of U.S. Cobra helicopter gunships overhead. Below, Iraqi fighters were trying to reach the abandoned track, with its load of weapons and ammunition. "We're hitting them," Martin recalled, "watching them drop."

He remembers a strange sensation. "Your body and brain ain't working like a normal person's would. Some people will snap. Some people will go off the edge. Everyone reacts differently," Martin said. "I was having fun."

Marines at the south bridge had picked up the radioed pleas for help and organized a rescue party. The first vehicle to arrive was a Humvee carrying a grizzled gunnery sergeant from another company. He was firing a pump-action shotgun out the passenger window as Marines on the roof sprayed cover fire.

"What do you need?" he shouted.

Water and radio batteries, the Marines answered.

"I'll be back," the Gunny said.

An M1-A1 tank arrived soon after and took away the wounded. The gunnery sergeant then returned with Humvees to rescue the remaining Marines. As the vehicles unleashed heavy fire in several directions, forcing the Iraqi fighters to take cover, the Gunny stepped onto the street and lit a cigarette.

"God, I hate this place," he said. "Let's get the hell out of here."

Part of Charlie Company was still pinned down at the north bridge. Lance Corporal Killeen, the Florida weightlifter, and his platoon were in the swamp near the span. He could hear enemy soldiers nearby.

"I thought they were going to sandwich us," Killeen said. "I figured my life was all over."

It was nearing 1600 and the sun was getting low. The ground rumbled and Killeen climbed toward the bridge. If Iraqi tanks were coming, then the company was almost certainly lost, he recalled thinking. As the tanks neared the canal, he saw they were American. These were the tanks that had spent their fuel retrieving members of the Army maintenance company that had been ambushed. Refueling had taken longer than expected because the pumps malfunctioned and it had to done by hand. When they learned that Charlie Company was taking a pounding, the tank crews cut short the refueling and rushed back to Nasiriyah. Now they were firing their 120-millimeter guns at Iraqi positions.

"It was the best feeling in the world," Killeen recalled.

As the Marines later prepared for a memorial service at Camp Lejeune, many were trying to recover from wounds - both mental and physical. Glass had eleven surgeries to remove shrapnel, dead muscle and metal pins from his leg after being wounded by an Iraqi RPG. His fibula was removed, and he had several skin grafts.

Martin had eight chunks of shrapnel from the A-10 removed from his hands and legs on the battlefield. Two additional pieces were removed from his arm on the Navy ship returning him to North Carolina.

Torres, who was also hit by the A-10, walked with the help of crutches as he recovered from shrapnel wounds to his right leg and left side. After being released from the Bethesda Naval Hospital he worked to regain movement in his left foot.

On March 23, 2003, the fourth day of the Iraq war, eighteen Marines died fighting to take a bridge in Nasiriyah. Nine served in a mortar squad that came under intense Iraqi bombardment.

Sergeant Michael E. Bitz,of Ventura, CA drove an amphibious assault vehicle, was wounded helping injured Marines, and was killed by an Iraqi shell.

Lance Corporal Thomas A. Blair of Broken Arrow, OK, was part of an air-defense team. He disappeared in the fighting and was later confirmed as killed in action.

Lance Corporal Brian R. Buesing of Cedar Key, FL, was in the mortar squad. His grandfather served in the same squad in the Korean War and won a Silver Star.

PFC Tamario D. Burkett of Buffalo, NY was a poet, an artist and the oldest of seven children. He was with the mortar squad.

Corporal Kemaphoom A. Chanawongse of Waterford, CT, a Thai immigrant, was a crew commander. He was hit by artillery fire while trying to retrieve ammunition.

Lance Corporal Donald J. Cline Jr. of Sparks, NV, was a rifleman. He said he was going to help wounded Marines and was not seen again. He was later confirmed dead.

Lance Corporal David K. Fribley of Fort Myers, FL, joined the service after the September 11 terrorist attacks. He was killed by friendly fire from an Air Force A-10 fighter.

Corporal Jose A. Garibay of Costa Mesa, CA, a Mexican immigrant, was part of the mortar squad. A shell destroyed a vehicle evacuating him and other wounded Marines.

Private Jonathan L. Gifford of Decatur, IL, an outdoorsman, was a member of the mortar squad.

Corporal Jorge A. Gonzalez of El Monte CA, wanted to become a police officer. He was with the mortar squad.

Private Nolen R. Hutchings of Boiling Springs, SC, enlisted in the Marines after high school. He was with the mortar squad.

Staff Sergeant Phillip A. Jordan of Enfield, CT had served fifteen years in the Marines. He was with the mortar squad.

2nd Lieutenant Frederick E. Pokorney Jr. of Tonopah, NV was a forward artillery observer. He died trying to call in artillery strikes on Iraqi positions.

Lance Corporal Patrick R. Nixon of Gallatin, TN came from a family whose members had served in every major conflict since World War I. He was with the mortar squad.

Sergeant Brendon C. Reiss of Casper, WY was a squad leader who had recently reenlisted. He was running to get more ammunition when he was hit.

Corporal Randal K. Rosacker of San Diego was a machine gunner who was providing cover fire after the Marines crossed the bridge. He was one of the first Americans killed.

Lance Corporal Thomas J. Slocum of Thornton, CO was in the hatch of a vehicle taking wounded Marines to the rear when he was hit.

Lance Corporal Michael J. Williams of Phoenix AZ gave up a flooring business to join the Marines. He was with the mortar squad.

This story appeared in the Los Angeles Times on August 26, 2003.

A PROUD FIGHTER PILOT

"The thing I'm most proud of is that I was a Marine Corps fighter pilot." - Ted Williams

Anyone who knows me knows I am a lifelong Yankee fan, and that I hate the Boston Red Sox worse than communism. However, I do admit I admire the loyalty of the long-suffering fans of Boston, partly because my dear departed Mother was one of them - but I hate the team nonetheless.

I like to draw a parallel between the New York Yankees and the United States Marine Corps, two organizations which have had a major impact on my life. Each of them is a brotherhood built upon sustained excellence, rich traditions, and the bottom line - winning. Think about it. The Yanks have twenty-six Championships, which is the most in professional sports, and the Corps has an unmatched battle record. The Yanks had Ruth, Gehrig, DiMaggio and Mantle, and the Corps had Puller, Basilone, Daly and Butler. The Yanks play in the baseball cathedral known as Yankee Stadium, and the Marine Corps conducts the Evening Parade on the Commandant's lawn at the oldest post of the Corps at Eighth & I. Yankee players donning the venerable pinstripes feel a sense of reverence in doing so, and Marines take the same pride in wearing the Eagle, Globe and Anchor. The similarities just go on and on.

It is not an accident the "also-rans" in each case, be it the lesser teams of baseball or the lesser military services, are jealous of this distinction, and harbor feelings of ill will toward the giant in their arena. Fans of other teams charge the Yankees with unfairly dominating the sport of baseball in every way imaginable. They just don't understand the Yankee mystique. And in a similar fashion, members of our sister services try to belittle the Corps at every opportunity. I guess they figure it is easier to tear down the best than to raise themselves to a higher level. I suppose that is the price we must pay for being the best.

With all of that said I firmly believe, in spite of my undying allegiance to the Yankees and unmitigated disdain for the Red Sox, that the greatest hitter of all time played for Boston. As you can imagine it pains me greatly to say that, and the only thing that gives me any solace is the fact he was also a United States Marine.

As I am sure you have guessed, the man I am talking about is Ted Williams. Much has been said about his records - the lifetime .344 batting average, the 521 career home runs, and of course his feat of hitting .406 in 1941. He did all of that despite missing five seasons in the prime of his career while serving his country during World War II and Korea, and many believe he would have erased every hitting mark in the record book if that had not been the case. I tend to agree.

Much has been written about Ted Williams the ballplayer, and I am not going to attempt to duplicate that here. Not nearly enough has been said about Ted Williams the Marine. While in Korea, Williams flew thirty-nine missions. He later downplayed his combat record, writing: "I was no hero. There were maybe seventy-five pilots in our two squadrons and ninety-nine percent of them did a better job than I did." But his record is nothing to make light of. He served our country with distinction, when others might have resisted the call. He did his duty.

There is a famous photograph of Williams standing next to a recruiting poster bearing his likeness, and what he did in that picture speaks volumes about his pride in being a

Marine. The words on the poster say "He Was A Marine," and Williams is shown holding a hand-lettered card up to the poster that said, simply,"IS." After all, once a Marine, always a Marine.

Another telling event occurred in the wake of Ted Williams' death. He had lived the last years of his life in Winter Park, Florida, where he constructed a museum honoring the greatest hitters in baseball history. When the news broke Williams had succumbed to his illness, there was a story on the local news about an impromptu tribute on the steps of that museum, not by Red Sox fans, but by Marines. They had gone there in part to pay homage to a legend, but mostly they wanted to honor the service and sacrifice of a brother Marine.

The article below, written by Joe Burris the day after Williams was honored in a memorial service at Fenway Park, gives the views of one of those who knew Ted Williams best - Senator John Glenn.

As for me, I am proud to say Ted Williams was a Marine - but I will always believe he would have looked great in Yankee pinstripes:

When the United States Marine Corps fighter pilot, astronaut and senator who flew missions over Korea with Ted Williams heard that Williams had complained about serving his country at that time - which was his second military stint - John Glenn sought to shoot such perceptions out of the sky.

"I never heard Ted complain one time," said Glenn, who participated in a memorial service at Boston's Fenway Park to honor the man he calls the best wingman he ever flew with.

Glenn said the two flew half of Williams' thirty-nine missions together, as part of the BMF-311 Squadron. He added that though Williams is known as the best hitter who ever lived, his fondest memories of the Red Sox slugger are of an excellent Marine fighter pilot who served his country proudly.

"I don't know what kind of records he might have set had he not been called back in. Who knows what would have happened with the record book if he had that time in baseball? But he never complained once, and I don't think there was ever anyone prouder to be a Marine than Ted Williams."

"I think Ted's a good example of the attitude people should have toward serving their country. I never heard Ted gripe once - not once - about being called back in, even though some of his best years in baseball might have occurred at that time. Not a single word. He did his job. He was good. He concentrated on it. He was dedicated to being just as great a pilot as he was a ballplayer. And he was a good one."

Glenn relayed that to the Fenway crowd, and when he said Williams didn't complain, "not one word" about serving, the crowd applauded. He also told a story about Williams being spotted in Japan by a group of children who wondered if he was the major league player.

Neither could speak the other's language, so one of the youngsters got into a batting stance, swung, and pointed to him, and Williams laughed and nodded to say he was indeed the great major leaguer.

"The boy got into the stance again, but this time Ted scowled at him," said Glenn, much to the laughter of the crowd. "He went over to where the boy was standing, put him in a batting position, and proceeded to correct his form," added Glenn. "He made this little kid stand with his feet farther apart, more bent at the waist, head more lined up with an imaginary pitcher, and arms farther back to start his swing. "Leaving the boy in that stance, Ted stepped back several paces and took a pitching position and threw an imaginary baseball toward the boy, who swung with all his might. Ted ducked, let out a whooshing, whooping big yell, and swung around 180 degrees as though that non-existent baseball was headed to the center field bleachers. The kids were jumping up and down, and no one was more surprised than the batter – I'm sure he was an instant hero. Ted got a bigger kick out of that than the kids did. As a spoken language, not a word was understood in either direction. But it was not necessary."

The first American to orbit the earth, John Glenn flew ninety combat missions in the Korean War and shot down three enemy planes. He also served as a Marine Corps pilot in World War II, where he earned five Distinguished Flying Crosses and nineteen air medals.

He said he met Williams while Williams was taking a jet refresher course in North Carolina in 1952, just before the two were shipped out to Korea. He spoke of a pilot who was as fearless in the skies as he was at the plate. He also spoke about a combat incident that almost cost Williams his life.

"He got hit by anti-aircraft fire and had an engine fire coming out the back of his airplane," said Glenn. "Usually on the old F9F, the Panther, when that had happened on previous flights with somebody, they would have to bail out or the tail blew off the airplane. Ted didn't want to get out.

"His radio was out. His hydraulics were out. He came around and made a belly landing because he couldn't get the landing gear down. He slid up the runway about 1,500 feet, got out, and watched the plane burn. He was about as close as you get on that one of not surviving."

Glenn also spoke of the time when Williams was flying in low while they were making an attack on a fortified bunker and another bunker exploded beneath him. He said an eight-inch rock flew up into Williams' fuel tank and made a jagged hole in it.

"He was an excellent fighter pilot and he wasn't one that hung back, he was in there pressing the attacks like everyone else," said Glenn. "I remember one time someone asked him what was his favorite music, and he said without hesitation, 'The Star Spangled Banner, the Marines Hymn, and Take Me Out to the Ballgame.'"

John Glenn retired from the astronaut program in 1964 and won election to the Senate in 1974, then was re-elected from 1980-92. When he failed to win the 1984 Democratic presidential nomination he decided not to run for Senate re-election in 1998.

Glenn said that among his more animated conversations with Williams were ones about politics, particularly since Williams was a Republican. "We had some great political discussions. He was not one to hold back on his views, and neither was I."

When John Glenn returned to space aboard the shuttle Discovery in 1998 he was seventy-seven years old, making him the oldest person ever to participate in space travel. He invited Williams to come to Cape Canaveral for the launch, and Williams obliged.

"Ted's son brought him over in a wheelchair," said Glenn. "Of course I didnt see this, but I was told that when I was taking off and the rocket was going off and clearing the pad, Ted stood up out of his wheelchair and said, 'That's my friend.'"

This story appeared in the Boston Globe on July 23, 2002.

CAPTAIN PARKER

By Charles R. "Chuck" Dowling

"Marines divide the world into two classes: Marines, and those who aren't good enough to be Marines." – David B. Wood

As new Lieutenants, the courses we took at Basic School back in the "Old Corps" included an introduction to hand-to-hand combat. As expected we went through all the "trip and flip" techniques, inspired by what was known then as jiu-jitsu. We also received instruction on killing with bare hands, knives and axes. Where we would get an axe in combat was not made clear to us. The instructor was a Marine Captain named Parker, and he was the most glorious specimen of mankind I had ever seen. At six feet four inches and about two hundred and thirty pounds of compact muscle, he was the epitome of the perfect man. A modern day sports fan would say he had the build of the ideal outside linebacker.

As I remember the course, it was quite graphic and gruesome. We learned how to strangle someone with our hands, and how to use a garrote to cut into or even slice off a man's head. He showed us how to disembowel an enemy with a knife or bayonet, to crush heads with a rifle, and how to kill a man by hitting him in the back with an axe in order to sever the spinal cord.

We also learned that Captain Parker was a humorless and focused individual. His pale blue eyes were clear and menacing. His face, although handsome and even boyish looking, was hard and unforgiving. We came away from the course understanding we had just met someone who was, to use a cliché, one of a kind and "one bad dude."

After Basic School I went one way and Parker went another, but a few years later I bumped into some Lieutenants I had gone through basic school with and the name came up again. They had recently returned from participating in some landing operation exercises in the Philippines, and over a few cold ones at the "O" club we celebrated our reunion and exchanged a number of "sea" stories. One of them was about Captain Parker.

It seems that Parker, as part of a large scale training exercise in the Philippines, had "jumped" onto a small island at the edge of the area of operations with a small Force Recon team. The jungle island, some ten miles off the coast, was uninhabited, and for the purposes of the exercise represented an "aggressor" radar site. Their mission was to neutralize the facility and to set up their own communication system. They were also to provide early warning of any "aggressor" activity on the flank of the operation.

The great God Murphy, of Murphy's law fame, brought his evil brothers with him for this exercise. The main landings were a mass of confusion. Many units landed in the wrong places and missions had to be changed or scrapped. (This is why we practice, was the face saving wisdom). A few real life accidents occurred where some Marines were injured, stopping parts of the exercise completely, and as if that wasn't enough on the third day a totally unpredictable (so the weather people said) tropical storm swept into the area. The storm forced a number of the attack and support vessels to move to open water to gain maneuver room.

After the storm passed the exercise was continued with little additional trouble. At the end of ten days, the operation completed, the Navy and the Marines packed up and left for home. The original time period for the exercise was to have been five days, but the mishaps and the storm had added an extra five.

Meanwhile Captain Parker and his team had supplies for three days, and were to have been picked up or re-supplied at the end of that time. As with the main "invasion," one of Murphy's miserable minions spread his largesse on Parker and his men too. Their radio went down on the second day and was not fixable, and due to the problems the main effort was encountering the required twice-daily reports from Parker were not missed. As a matter of fact, Parker and his team were completely lost in the shuffle.

The retreating armada was two days at sea when someone realized Parker had been left behind. An LST was sent back and three days later they were picked up, alive, hungry and more than a little angry. Force Recon people know there will be times when they have to eat snakes and bugs, but they weren't anticipating it happening on a training exercise.

As the story goes Captain Parker took his men to the galley and demanded they be fed immediately, and the mess crew fell all over themselves to be accommodating. Parker then went to the officer's wardroom, slammed the door open, shouted for food and sat down at a table set with a white tablecloth, white linen, monogrammed silverware, crystal, and bone china.

He had lived on the ground and in the mud for more than two weeks. He had worn the same clothes for more than two weeks. He had not bathed or shaved for fifteen days. For two weeks he been drinking polluted water, which had caused dysentery and left him smelly and a little emaciated. He had not eaten food for twelve days. At that moment he was not the most rational or squared away Marine you will ever see. He was also not properly attired for the wardroom.

The stewards jumped to his command however. Dinner was about to be served to the first shift of officers, my friends among them, so they were ready. One of the stewards grabbed a full plate and a basket of bread and put it in front of Parker. Just at that moment the officer in charge of the mess, a Naval Lieutenant, entered the wardroom, took one look at Parker and the filth and the dirt he had trailed in, and ordered him out.

Parker took some of the meat off the plate with his filthy hands, rolled it up into a piece of bread he had taken from the breadbasket, and took a big bite without even looking at the lieutenant.

The Lieutenant ordered him to leave again, this time with his voice at a higher pitch. Parker did not even acknowledge his existence. He made himself another sandwich and continued to eat. The Lieutenant's face was reddening. By then some of the other officers sensed trouble, and were attempting to get the Lieutenant to calm down. He refused to listen, ranting about Marines with no decorum and dignity and flaunting inappropriate behavior for an officer etc. He didn't care what the situation was, etc. etc. No one comes into the wardroom like that, etc., etc…

Breaking free from those attempting to calm him, he leaned over the table, facing Parker. He lowered himself so he was right in his face, and had both hands splayed in front of him and resting on the table. "Get out!" he ordered as loudly as he could.

Those who were there said they had never seen hands move as fast as Parker's hands moved. Within what appeared to be split seconds, Parker had pulled his combat knife and slashed down at the table in a lightening burst of quick jabs. When he finished, barely moments from when he pulled the knife, there were slices in the tablecloth between each of the

Lieutenant's ten fingers - and the knife was stuck in the table in front of the grubby Marine, still quivering. He had, with great skill, missed every one of the Lieutenants fingers by the smallest of margins. He then made himself another sandwich, licking a small spill of gravy off his dirty fingers.

"Good meat," he said as he smiled up at the Lieutenant. The Lieutenant had not moved. He just looked at Parker in ashen-faced horror.

In the end the Captain of the LST, a senior Lieutenant, was a reasonable man. Parker agreed to pay the officers mess for a new tablecloth, and the issue was settled.

My buddies, the Lieutenants telling me the story, said they saw Parker off and on for the next few days as he strolled around the vessel, but they did not see the Naval Lieutenant for the rest of the trip.

BAA BAA BLACK SHEEP

"Just name a hero, and I'll prove he's a bum."
– Colonel Gregory "Pappy" Boyington

We Marines have always prided ourselves on being squared away, positive role models. We have the Silent Drill Platoon. We spit shine our boots. We field day the barracks until you can eat off the deck, and stand wall locker inspections where one "Irish pennant" is cause for failure. But the thing we admire above all else is personal heroism.

I think many of us live vicariously through larger than life "bad boy" characters like Pappy Boyington. We would never pull some of the stunts he did, and in fact a lot of his conduct would be cause for dismissal from the modern day Marine Corps. But when combat is involved the niceties of peacetime get cast aside, and characters like Boyington tend to flourish.

I don't watch a lot of television, but when I do I enjoy much of the programming on the History Channel. I have seen a number of documentaries about the Marine Corps there, and particularly enjoy the show "Black Sheep Squadron." While much of that show is pure "Hollywood" and must be taken with a grain of salt, many of the incidents are loosely based upon fact, and good use is made of actual aerial combat footage. But the best part is the interviews with the actual pilots of VMF-214, the "Black Sheep". These are the guys who actually flew with Boyington, and listening to them reminisce is worth tuning in all by itself.

While I certainly don't recommend emulating Pappy Boyington during a career in the Marine Corps, his story sure is interesting. If another big war comes along all bets will probably be off once again, but until that day we will have to content ourselves with reading about his exploits:

Undoubtedly the most colorful and well known Marine Corps ace was Gregory "Pappy" Boyington, Commanding Officer of VMF-214. Stories about Boyington are legion, many founded in fact, including how he led the legendary Black Sheep squadron and served in China as a member of the American Volunteer Group - the famed Flying Tigers. He spent a year and a half as a Japanese POW, was awarded the Medal of Honor, and was recognized as the Marine Corps' top ace. Always hard-drinking and hard-living, Pappy's post-war life was as turbulent as his wartime experiences.

Born on December 4, 1912, young Greg had a rough childhood - divorced parents, alcoholic step-father (whom Greg believed to be his natural father until he entered the Marine Corps), and lots of moves. He grew up in St. Maries, Idaho, a small logging town, and got his first ride in an airplane when he was only six years old. The famous barnstormer, Clyde Pangborn, flew his Jenny into town, and Greg wangled a ride. What a thrill for a little kid! Greg's family then moved to Tacoma, Washington in 1926. While there in high school he took up a sport that he would practice for many years - wrestling. When he'd had a few too many drinks (which was often), adult Boyington would challenge others to impromptu wrestling bouts, frequently with injurious results. He enrolled at the University of Washington in 1930, where he continued wrestling and participated in ROTC. He met his first wife Helene there, and they were married not long after his graduation in 1934. His first son, Gregory Clark Boyington, was born ten months later.

After a year with Boeing, Greg enlisted in the Marine Corps. It was then, when he had to supply them with his birth certificate, that he learned of his natural father. He then began elimination training in June of 1935, where (in the small world of Marine aviation at that time) he met Richard Mangrum and Bob Galer, both future heroes at Guadalcanal. He passed, and received orders to begin flight training at Pensacola NAS in January of 1936 with class 88-C. There he flew a floatplane version of the Consolidated NY-2. Like another great ace, Gabby Gabreski, Boyington had a tough time with flight training and had to undergo a number of rechecks.

Until he arrived in Pensacola, Boyington had never touched alcohol. But there, with hard-partying fliers and aware of his wife's "fooling around," he soon discovered his affinity for liquor. Early on, Boyington established his Marine Corps reputation: hard-drinking, brawling, well-liked, and always ready to wrestle at the drop of a hat. He kept flying all through 1936, slowly progressing toward earning his wings and flying more powerful planes like the Vought O2U and SU-1 scouting biplanes. At Pensacola he also met his future nemesis, Joe Smoak, memorialized in *Baa Baa Black Sheep* as "Colonel Lard." He finally won his coveted wings in March of 1937, becoming Naval Aviator #5160.

Before reporting for his assignment with VMF-1 at Quantico, Virginia Boyington took advantage of his thirty-day leave to return home and reconcile with his wife Helene, who became pregnant with their second child. In those days Marine aviators were required to be bachelors, so Greg's family was a secret that he kept from the brass - but he brought them with him to Virginia and installed them quietly in nearby Fredericksburg. He flew F4B-4 biplanes during

1937, taking part in routine training, an air show dubbed the "All American Air Maneuvers," and a fleet exercise in Puerto Rico.

In July of 1938 he moved to Philadelphia to attend the Marine Corps' Basic School for ten months. Apparently not motivated by the "ground-pounder" curriculum, Boyington here evidenced the weaknesses that would haunt him: excessive drinking, borrowing money and not repaying it, fighting, and poor official performance.

His irresponsibility, his debts, and his difficulties with the Corps continued to mount throughout 1939 and 1940 when he flew with VMF-2 while stationed at San Diego. One memorable, drunken night he tried to swim across San Diego Bay, and wound up naked and exhausted in the Navy's Shore Patrol office. Despite his problems on the ground, it was during these days of 1940 while flying with VMF-2 that Boyington first began to be noticed as a top-notch pilot. Whatever his other issues, he could out-dogfight almost anyone - but back at Pensacola in January of 1941 his problems mounted when he decked a superior officer in a fight over a girl, and his creditors sought official help from the Marine Corps. Greg's career was a hopeless mess by late 1941.

Rescue came from, of all places, China. Anxious to help the Chinese in their war against Japan, the U.S. government arranged to supply fighter planes and pilots to China under the cover of the Central Aircraft Manufacturing Company (CAMCO). CAMCO recruiters visited U.S. military aviation bases looking for volunteers. As Bruce Gamble described it in *Black Sheep One*:

"The pilots were volunteers only in the sense that they willingly quit their peacetime job with the military; otherwise they were handsomely paid through CAMCO.

Pilots earned $600 a month and flight leaders $675, plus a fat bonus for each Japanese plane destroyed. This was double or even triple the current military salary for pilots... in March, CAMCO representatives began recruiting military pilots for what would become the American Volunteer Group (AVG)... one recruiter set up an interview room in San Diego's San Carlos Hotel, a popular watering hole for pilots, and on the night of August 4 Greg Boyington found himself in the hotel bar simply 'looking for an answer.' Payday had been just a few days earlier, but already he was broke. His wife and children were gone, he was deeply in debt, and many of his superiors were breathing down his neck."

The money looked very good to Boyington. Assured that the program had government approval and that his spot in the Corps was safe, he signed on the spot and promptly resigned from the Marine Corps. While the AVG deal for pilots normally did contemplate a return to active U.S. military service, in Greg's case his superiors took a different view. They were happy to be rid of him, and noted in his file that he should not be reappointed.

He shipped out of San Francisco on September 24, 1941 aboard the *Boschfontein* of the Dutch Java Line. After docking at Rangoon, the AVG fliers arrived at their base at Toungoo on November 13th. Boyington flew several missions during the defense of Burma, and after Burma fell returned to Kunming and flew from there until the Flying Tigers were incorporated into the USAAF. His autobiography includes many war stories from his experiences with the Flying Tigers.

Boyington clashed with the leader of the Flying Tigers, the strong-willed Claire Chennault, and quit the AVG in April of 1942. Chennault gave him a dishonorable discharge, and Greg went back to the U.S. While there he claimed to have shot down six Japanese fighters, which would have made him one of the first American aces of the war. He maintained until his death in 1988 that he did in fact have six kills, and the Marine Corps officially credits him with those victories.

While with the Flying Tigers Greg also made the acquaintance of Olga Greenlaw, the XO's beautiful wife who, in her own words, "knew how to get along with a man if I liked him." Apparently she and Boyington "got along." She even wrote her own book, *The Lady and the Tigers*, in 1943.

Boyington subsequently returned to the States in the spring of 1942 and took up with a woman named Lucy Malcolmson - his first marriage having fallen apart. Then with some finagling, undoubtedly helped by the wartime demand for experienced fighter pilots, he was reappointed to the Marines in November with the rank of Major. In January of 1943 Boyington embarked on the *Lurline*, bound for New Caledonia, where he would spend a few months on the staff of Marine Air Group (MAG)-11. Here he got his first close look at a Corsair, flown by his friend Pat Weiland.

Boyington finally secured assignment to VMF-122 as Executive Officer for a combat tour, but as usual he clashed with his superior - this time Major Elmer Brackett. In any event Brackett was shortly removed and Boyington took over, but did not see much action. It was at this time, in early 1943, when as the new CO of VMF-122 his claim of six kills with the AVG first made it into print. Then Smoak relieved him of his command of VMF-122 in late May, and that was followed by a broken leg and time in the hospital.

In the summer of 1943, as Boyington convalesced, the U.S. naval air forces needed more Corsairs in the fight.

Oddly, the key pieces - trained pilots and operational aircraft - were present in the South Pacific, but many of them were dispersed. Who got the idea remains unclear, but he was given the assignment to pull together an ad hoc squadron from available men and planes. Originally they formed the rear echelon of VMF-124, but eventually these twenty-six pilots would become the famous “Black Sheep.” In a complex, and common, wartime shuffling of designations, Boyington’s team was redesignated VMF-214, while the exhausted pilots of the original VMF-214 (nicknamed the Swashbucklers) were sent home.

Under Boyington as CO and Major Stan Bailey as Exec, they trained hard at Turtle Bay on Espritu Santo, especially the pilots who were new to the Corsair. While leading this group of young pilots, most in their early twenties, Boyington - at the “advanced” age of thirty - picked up the nickname “Gramps.” The Black Sheep don’t remember calling him “Pappy”- that was a nickname the press picked up after he was shot down.

In early September of 1943 the new VMF-214 moved up to their new forward base in the Russells, staging through Guadalcanal’s famed Henderson Field. The “Black Sheep” fought their way to fame in just eighty-four days, piling up a record 197 planes destroyed or damaged, troop transports and supply ships sunk, and ground installations destroyed in addition to numerous other victories. They flew their first combat mission on September 14, 1943, escorting Dauntless dive bombers to Ballale, a small island west of Bougainville where the Japanese had a heavily fortified airstrip. There they encountered heavy opposition from enemy Zeros. Two days later, in a similar raid, “Pappy” claimed five kills, his best single day total. In October VMF-214 moved up from their original base in the Russells to a more advanced location at Munda. From here they were closer to the next big objective - the Jap bases on Bougainville. On one mission over Bougainville, according to Boyington's autobiography, the Japanese radioed him in English, asking him to report his position and so forth. Pappy played along, but stayed five thousand feet higher than he had told them, and when the Zeros came along the Black Sheep blew twelve of them away.

During the period from September 1943 to early January 1944 Boyington destroyed twenty-two Japanese aircraft. By late December it was clear he was closing in on Eddie Rickenbacker's record of twenty-six victories (including his six with the AVG), and the strain was starting to tell. On January 3, 1944 Boyington was shot down in a large dogfight in which he claimed three enemy aircraft, and was captured. The following is an excerpt from Boyington's book *Baa Baa Black Sheep* describing his final combat mission:

"It was before dawn on January 3, 1944, on Bougainville. I was having baked beans for breakfast at the edge of the airstrip the Seabees had built, after the Marines had taken a small chunk of land on the beach. As I ate the beans, I glanced over at row after row of white crosses, too far away and too dark to read the names. But I didn't have to. I knew that each cross marked the final resting place of some Marine who had gone as far as he was able in this mortal world of ours.

Before taking off everything seemed to be wrong that morning. My plane wasn't ready and I had to switch to another. At last minute the ground crew got my original plane in order and I scampered back into that. I was to lead a fighter sweep over Rabaul, meaning two hundred miles over enemy waters and territory again. We coasted over at about twenty thousand feet to Rabaul. A few hazy cloud banks were hanging around - not much different from a lot of other days. The fellow flying on my wing was Captain George Ashmun from New York City. He had told me before the mission: 'You go ahead and shoot all you want, Gramps. All I'll do is keep them off your tail.'

This boy was another who wanted me to beat that record, and was offering to stick his neck way out in the bargain. I spotted a few planes coming through the loosely scattered clouds and signaled to the pilots in back of me: 'Go down and get to work.' George and I dove first. I poured a long burst into the first enemy plane that approached, and a fraction of a second later saw the Nip pilot catapult out and the plane itself break out into fire. George screamed out over the radio: 'Gramps, you got a flamer!'

Then he and I went down lower into the fight after the rest of the enemy planes. We figured that the whole pack of our planes was going to follow us down, but the clouds must have obscured their view. Anyway, George and I were not paying too much attention, just figuring that the rest of the boys would be with us in a few seconds, as was usually the case. Finding approximately ten enemy planes, George and I commenced firing. What we saw coming from above we thought were our own planes - but they were not. We were being jumped by about twenty planes. George and I scissored in the conventional Thach weave way, protecting each other's blank spots, the rear ends of our fighters. In doing this I saw George shoot a burst into a plane and it turned away from us plunging downward, all on fire. A second later I did the same thing to another plane. But it was then that I saw George's plane start to throw smoke, and down he went in a half glide. I sensed something was horribly wrong with him. I screamed at him: 'For God's sake, George, dive!'

Our planes could dive away from practically anything the Nips had out there at the time, except perhaps a Tony. But apparently George had never heard me or could do nothing about it if he had. He just kept going down in a half glide. Time and time again I screamed at him, but he didn't even flutter an aileron in answer to me.

I climbed in behind the Nip planes that were plugging at him on the way down to the water. There were so many of them I wasn't even bothering to use my electric gun sight consciously, but continued to seesaw back and forth on my rudder pedals, trying to spray them all in general, trying to get them off George to give him a chance to bail out or dive - or do something at least. But the same thing that was happening to him was now happening to me. I could feel the impact of enemy fire against my armor plate, behind my back, like hail on a tin roof. I could see the enemy shots progressing along my wing tips, making patterns.

George's plane burst into flames and a moment later crashed into the water. At that point there was nothing left for me to do. I had done everything I could. I decided to get the hell away from the Nips. I threw everything in the cockpit all the way forward - this means full speed ahead - and nosed my plane over to pick up extra speed until I was forced by water to level off. I had gone practically a half a mile at a speed of about four hundred knots, when all of a sudden my main gas tank went up in flames in front of my very eyes. The sensation was much the same as opening the door of a furnace and sticking one's head into the thing.

Though I was about a hundred feet off the water, I didn't have a chance of trying to gain altitude. I was fully aware that if I tried to gain altitude for a bail-out I would be fried in a few more seconds."

Boyington landed in the water, badly injured. After being strafed by the Jap fighters, he struggled onto his raft until captured by a Jap submarine several hours later. They took him first to Rabaul, where he was brutally interrogated. Even the general commanding Japanese forces at Rabaul interviewed him. Pappy related in *Baa Baa Black Sheep* that the general asked him who had started the war. After Pappy replied that of course the Japanese had started the war by attacking Pearl Harbor, the general then told him this short fable:

"Once upon there was a little of old lady and she traded with five merchants. She always paid her bills, and got along fine. Finally the five merchants got together, and they jacked up their prices so high the little old lady couldn't afford to live any longer. That's the end of the story."

The general left the room, leaving Boyington to ponder that there had to be two sides to everything.

After about six weeks the Japanese flew him to Truk, where he experienced one of the early carrier strikes against that island in February of 1944. Along with six other captured Americans, he was confined in a small but sturdy wooden cell - which might have been designed for one inmate. The only opening was a six-inch hole in the floor for relieving themselves. With six men in a tiny cell this was unpleasant enough, but when the Japs actually overfed them with rice balls and pickles diarrhea resulted, and then the situation became really messy.

He eventually moved to a prison camp at Ofuna, outside of Yokohama. His autobiography relates the frequent beatings, interrogations, and near starvation that he endured for the next eighteen months. The guards, whose only qualification seemed to be passing "a minus-one-hundred

I.Q. test," beat the prisoners severely for any infraction, real or imagined.

Boyington lost about eighty pounds, and described how he once entirely consumed a "soup bone the size of my fist" in just two days, a feat which previously he would not have believed a dog could achieve. During the middle period of his captivity he had the good fortune to be assigned kitchen duty. Here, a Japanese grandmother who worked in the kitchen befriended Greg and helped him filch food. Before long, he returned to his pre-captivity weight. He even got drunk on New Year's Eve, begging a little sake from each of the officers. From Camp Ofuna, he witnessed the first B-29 raids striking the nearby naval base at Yokohama.

When he was repatriated, Boyington found out he had been awarded the Medal of Honor and the Navy Cross. He also added to his claims for aerial victories after his return. Several other pilots had seen him down one Zero, which raised his total to twenty with the Black Sheep, and twenty-six if his six with the Flying Tigers were included. Twenty-six was Eddie Rickenbacker's WWI record, and also the number shot down by Joe Foss, the top-scoring Marine pilot of all time. When the final two kills he claimed before being shot down were added, the total rose to twenty-eight.

Pappy lived until 1988 but it was a hard life marked by financial instability, marriages, divorces and battles with alcoholism. However it has been noted by many that, whatever his problems, he never seemed to lack for attractive female companionship. Things started downhill on his War Bond tour, when he was frequently drunk, and the Marine Corps placed Boyington on the retired list in 1947, allegedly for medical reasons.

Of course Pappy's greatest fame came in the mid Seventies, when the television show *Baa Baa Black Sheep* debuted. Based very loosely on Boyington's memoirs, the show had a three-year run and achieved a consistent popularity in re-runs. Pappy was a consultant to the show and got on well with its star, Robert Conrad - but the show's description of the Black Sheep pilots as a bunch of misfits and drunks, which Pappy happily went along with, destroyed his friendship with many of the squadron veterans. The show made Pappy a real celebrity, and along with his fourth wife Jo he made a good career out of being an entertainer - appearing at air shows, on TV programs, etc. Finally, after a long battle with cancer, Pappy died in 1988 - and so ended one of the most colorful lives any of us could hope to live.

SAFE IN KANDAHAR

"Don't be a fool and die for your country. Let the other poor, dumb sonofabitch die for his." - General George S. Patton

The following is an excerpt from an e-mail home by an Air Force Ordnance Demolition Unit team-member describing his first night at Kandahar airport in Afghanistan. The other services don't always tell us to our face, but sometimes they DO appreciate having us around:

"One of the perimeter positions only a hundred yards or so to our left took some incoming fire and we all went to general quarters, taking defensive fighting positions in our bivouac in case they penetrated to our position. The Marines quickly repelled the attack. It will not bother me should I live my entire life without having to kill a man, but I have to say I'm glad to be surrounded by a thousand nineteen-year-old Marines who can't wait to. They will be leaving in a few weeks and turning things over to the Army. I will miss them.

The only tents the Marines use are one-man pup tents and they are everywhere. Each foxhole and DFP (defensive fighting position) around the camp is accompanied by two of these humble little tents.

I have a renewed respect for the Marines. They arrived a month ago, dug in, and have been living out of these ridiculously small, 5x5 tents ever since. No heat, no latrines, no showers, nothing but backpacks, weapons, helmets and flak vests, and lots of ammo. And they've been doing it every day. Four man teams at each position, two sleeping, two on watch. God bless them, every one."

RENDERING HONORS

Chaplain Joseph P. Carlos, USN

"Today, the world looks to America for leadership. And America looks to its Corps of Marines." – President Ronald Reagan

On Tuesday, July 8th, 2003 several Marines spoke so loudly at Marine Corps Base Quantico that a couple of hundred people were impressed. The Marines didn't say a word, show off any high tech training, or prove they could kill. They simply showed respect. What they did spoke volumes about their "core values."

On that warm July morning Captain Kenneth Butrym, CEC, United States Navy was being laid to rest. He was only forty-six years old and he died suddenly, leaving a wife and three children aged ten to nineteen years. Navy personnel and Marines gathered at the funeral home the evening before, where they prayed and shared stories about how this man had such an impact on their lives. Approximately two hundred and fifty people came to worship at the Marine Memorial Chapel to celebrate life, rather than death, on Tuesday. After Mass we headed towards the back gate enroute to Quantico National Cemetery, the same hallowed ground where Captain Butrym and his wife Mary had buried their three-day-old daughter back in 1991.

The mortuary directors put my car directly in back of the limo transporting the family. It was between 1130 and noon when the cortege wound its way toward the back gate, and as usual Marines were running during their lunch hour. In groups of two's and three's, these Marines - with their green t-shirts and shorts soaked in sweat - stopped their run, faced the road, and stood at attention as the hearse drove by! Create that picture in your mind. This did not happen only once. It wasn't a unique experience. When the Marines saw the hearse, bearing a service member they had never met who was on his way to his burial, they showed respect. Oncoming traffic, heading into the base, pulled over to the shoulder of the road and stopped! There were one or two vehicles that just kept going, but they were the exception.

At the cemetery, while waiting for everyone to exit their cars and come to the grave, I asked the family's CACO, "Did you see the Marines?"

He said, "We all commented on it, and asked what would have happened on a Navy base?" I think both he and I suspected the answer. Respect. Honor. Marines. If you were a runner that day or driving your car and pulled over - the family thanks you for showing respect and dignity. Your action spoke louder than any words ever could!

A BOTTLE OF RED

A Bottle of White

"A bottle of white, a bottle of red, perhaps a bottle of rosé instead…" - From the song "Scenes from an Italian Restaurant" by Billy Joel

Right or wrong, the consumption of alcohol has almost always been a part of military life. From the issue of grog rations aboard sailing ships to the "wetting down" parties held to celebrate promotions, those in uniform have found many reasons to feed the "wild hair that grows where the sun don't shine." Of course alcohol, like anything else, can be a bad thing when it is abused - but those stories can be told in another forum.

If my Mother were alive today she would be gratified to know I never touched a single drop of alcohol - not even beer - during my youth. It was not until I had turned eighteen and graduated from boot camp that I actually had my first taste of demon rum. When I arrived home on "boot" leave some friends took me out bar-hopping to celebrate, and I didn't know what I was getting myself into. I remember at one point we found ourselves at a Long Island watering hole called Solomon Grundy's, and as I would soon come to learn the sight of a newly graduated "slick sleeve" private in uniform tends to attract a lot of free drinks. One former Marine sitting at the bar, whom I had never met, sent over a new pitcher of beer to our table every time it got so we could see the bottom of the previous one. It turned out to be quite a night, and of course the next morning I found myself experiencing my first hangover - as I recall it was about an eleven on the Richter Scale. And of course I said those famous last words we have all uttered at one time or another - "never again!"

Despite that unfortunate escapade my true apprenticeship didn't begin until I reported to my first unit at Camp Lejeune, and it wasn't long thereafter that I became a combatant in the "Battle of J'ville" down on Court Street in the town of Jacksonville. For those who have never been stationed at Lejeune, Court Street is the adult entertainment district common to all military towns, and back in those days it was home to nothing but tattoo parlors, pawn shops, and about thirty different bars. I spent many a night prowling that street with my buddies, and while I admit my time would have been better spent doing almost anything else, the value of that education in friendship and camaraderie cannot be easily measured.

Fortunately I got all of that out of my system pretty quickly. I know there are some who never really do, and it haunts them throughout their lives. Over the ensuing years my consumption of alcohol was pretty much limited to those official occasions where one was almost expected to drink - Mess Nights, Birthday Balls, and wetting downs. It really wasn't until I went on the MSG program that I had an opportunity, or a reason, to drink with more regularity. The opportunity had to do with the location and nature of my assignments, as you shall see. The reason had a lot to do with my then-wife. She was driving me crazy!

During my tour of duty at my first post in the Congo I quickly discovered there was very little to do in the way of recreation. Brazzaville had no theaters, bowling alleys, sporting events or even television. It was up to us to find something to do, and not surprisingly our Marine detachment became the social focal point for the embassy community. The Marine House had a pool table, ping pong table, and even a full bar, and we regularly hosted events at which embassy personnel and our expatriate friends could relax and unwind. We also had a movie projector, and since we regularly received first run films as part of the Navy Motion Picture System we hosted a movie night once every week. It was on those occasions that we imbibed, drinking the potent local beer and enjoying one another's company.

An incident which stands out from my days in Brazzaville is the one which inspired this story's title. It was not uncommon for Americans living in Africa to come down with all sorts of exotic tropical maladies, and I was no exception. Near the end of my tour I developed an illness, probably as the result of eating the local food, which was beyond the ability of our Regional Medical Officer to diagnose or treat. As a result I was flown to the Army's Landstuhl Medical Center in Germany for tests and treatment, and aside from the tests themselves it turned out to be a pretty good trip - especially since it happened to be Oktoberfest.

As luck would have it I was required to wait a week between tests, and was left to my own devices during that time. When I tired of drinking the wonderful German beer and stuffing myself with schnitzel I decided to sample a bit of the grape, and signed up for a "wine probe" on the Rhine River - which was nothing more than a floating wine tasting. A Special Services bus transported a group of us to a picturesque little town on the river, and after a stroll through the village I boarded the boat for a departure scheduled for just before sunset. As I looked around for somewhere to sit I saw an empty seat at a table occupied by three attractive women, and they immediately waved me over and asked if I would like to fill the empty spot. Naturally, I obliged.

It turned out my new acquaintances were all nurses. One was an Army Captain stationed at Landstuhl (I fervently hoped she would not be involved in my upcoming colonoscopy), one was a Navy Reserve Lieutenant Commander from Colorado, and the third was a civilian friend of theirs. We ended up having a great time, and as the boat pulled into the dock at the end of the evening we all agreed it was much too early to call it a night. We had a long bus trip back to the base ahead of us, and there was only one thing that would make it tolerable - more wine!

The only problem was it was about half a mile to the village, and one of us would have to run there, buy some wine, and make it back before the bus departed. I'll give you one guess who was chosen to go. As soon as the gangway was in place I sprinted down the pier and headed for town, and I was determined not to come back empty handed. Upon getting there I eventually found a place to purchase some wine, but had to wait while they filled the bottles for me. I could hear the clock ticking. As soon as the transaction was complete I found myself running back toward the pier, with a bottle of red in my left hand and a bottle of white in my right. As I came around the last bend I could see the bus beginning to pull away in the distance, and couldn't decide whether to run faster or just sit down and start drinking. Fortunately one of the nurses spotted me, and moments later I saw the welcome sight of brake lights as the driver stopped the bus. When I boarded I was greeted by a chorus of cheers, and for the duration of the trip back to Landstuhl I was something of a minor celebrity. I can't say that was the best wine I have ever had, but it sure was the most deserved!

My next post, in Canberra Australia, had a completely different dynamic. There was plenty to do from a recreational standpoint, but even so the consumption of alcohol is the unofficial national sport in Australia. So it was not surprising that it was there I witnessed one of the most impressive drinking exhibitions I have ever seen.

One summer evening my detachment hosted a mess night at the Royal Military College Sergeant's Mess, and at the conclusion of the formal portion of the evening we adjourned to the bar as is customary. An Army Lieutenant with whom I was friendly was due to be promoted to Captain the following morning, and since both he and his Colonel were present I convinced them it would be fun to pin the new bars on right there in the Mess. The ceremony was brief and went off without a hitch - but that is when things got interesting.

An Australian Sergeant Major, who also served as the President of the Mess Committee (PMC), approached from the vicinity of the bar and presented the newly promoted Captain with what he liked to refer to as a "thermometer." It was a yard glass, which is in fact about a yard tall, filled with a bottle of port wine and a bottle of champagne. And it actually did look like a thermometer, since the heavier port had settled in the bottom.

The Sergeant Major then looked at his watch and informed Captain Ciccarelli that the Mess record for drinking one of those concoctions was sixty-eight seconds, and all of the Aussies present laughed because they knew he was joking. A thermometer was intended to be carried around and sipped over the course of the evening, not "skulled." But Ciccarelli didn't know that. As soon as he heard the so-called "record" his competitive juices began flowing, and without blinking an eye he immediately tipped up the huge vessel and began to suck down its contents. We knew he could never drink the whole thing, so the question became how far he would get before coming up for air. To a man we stood there in utter amazement - while he finished every drop!

John Ciccarelli is not a big man, so it wasn't long before he was passed out in a chair at the bar, and he later told me he spent the entire next day with his head on the desk in his office. I can only imagine how awful he must have felt. But in a few short moments he had become a legend.

I am not trying to glorify drinking, not by any means. I just wanted to point out that while the negatives associated with the consumption of alcohol are very real, there is an upside as well. To me the best part about drinking is when that normally stoic buddy of yours puts his arm around your shoulder in a drunken haze and slurs the heartfelt words, "I love you, man!"

CHRISTMAS IN THE CORPS

"Sleep well tonight – the Marines are on duty!"

The holiday season is often a difficult time of year for those in the military. I have walked a post on guard duty on many a holiday, and have been deployed for many more. I was even in boot camp for my first Christmas in uniform. I especially remember one season while I was stationed in Okinawa. As the holidays approached I recall browsing through the PX down at Camp Foster. There was Christmas music playing in the background, and it was quite pleasant until the song changed to Bing Crosby's "I'll Be Home for Christmas." A woman shopping nearby went berserk, and began to scream hysterically that she was NOT going to be home for Christmas for the second straight year. It was quite a scene.

On Christmas Eve those of us who could not afford to fly back to the States for the holidays, or who had to remain behind to hold down the fort, headed out the gate of Camp Hansen in search of some sort of diversion. With so many people gone on leave "Sinville" was pretty dead, and we ended up in a Japanese bar singing karaoke in a drunken stupor, which wouldn't have been so bad except they only had one song in English and we sang it over, and over, and over. To this day I can't hear "Yesterday" by the Beatles without cringing. Such wonderful Yuletide memories!

As Christmas approaches this year with Marines still serving in Iraq, Afghanistan and elsewhere, I thought it would be appropriate to include the true story behind a poem many of us have seen circulating for years, along with the inspiring text of that poem.

This piece, which sees wide circulation every Christmastime, is often credited to "a Marine stationed in Okinawa, Japan" (or, since 9-11, a Marine stationed in Afghanistan). Unfortunately, an Air Force Lieutenant Colonel named Bruce Lovely took credit for composing this poem, claiming he penned it on Christmas Eve 1993 while stationed in Korea - and he saw it printed under his name in the *Fort Leavenworth Lamp* a few years later:

"I arrived in Korea in July 93 and was extremely impressed with the commitment of the soldiers I worked with and those that were prepared to give their lives to maintain the freedom of South Korea. To honor them, I wrote the poem and went around on Christmas Eve and put it under the doors of U.S. soldiers assigned to Yongsan."

Regrettably, Lieutenant Colonel Lovely did a great disservice to his fellow servicemen by claiming authorship of "The Soldier's Night Before Christmas," because it had already been published in *Leatherneck Magazine* in December 1991, a full two years before Lovely supposedly wrote it. The *Leatherneck* version was titled "Merry Christmas, My Friend" and was attributed to James M. Schmidt, then a Lance Corporal stationed at Marine Barrack 8th & I in Washington, D.C.

According to Corporal Schmidt:

"The true story is that while I was a Lance Corporal serving as Battalion Counter Sniper at the Marine Barracks 8th & I, Washington, DC, I wrote this poem to hang on the door of the Gym in the BEQ. When Colonel Myers came upon it, he read it and immediately had copies sent to each department at the Barracks and promptly dismissed the entire Battalion early for Christmas leave. The poem was placed that day in

the Marine Corps Gazette, distributed worldwide, and later submitted to Leatherneck Magazine."

The following is Corporal Schmidt's version as printed in *Leatherneck*, which differs from the current Internet version in many places (particularly in Marine-specific wording that has since turned into Army references, and alterations in other places to maintain the line-ending rhyme scheme):

Merry Christmas, My Friend

'Twas the night before Christmas, he lived all alone,
In a one-bedroom house made of plaster and stone.
I had come down the chimney, with presents to give,
And to see just who in this home did live.

As I looked all about, a strange sight I did see,
No tinsel, no presents, not even a tree.
No stocking by the fire, just boots filled with sand.
On the wall hung pictures of a far distant land.

With medals and badges, awards of all kind,
A sobering thought soon came to my mind.
For this house was different, unlike any I'd seen.
This was the home of a U.S. Marine.

I'd heard stories about them, I had to see more,
So I walked down the hall and pushed open the door.
And there he lay sleeping, silent, alone,
Curled up on the floor in his one-bedroom home.

He seemed so gentle, his face so serene,
Not how I had pictured a U.S. Marine.
Was this the hero, of whom I'd just read?
Curled up in his poncho, a floor for his bed?

His head was clean-shaven, his weathered face tan.
I soon understood, this was more than a man.
For I realized the families that I had seen on that night,
Owed their lives to these men, who were willing to fight.

Soon around the Nation, the children would play,
And grown-ups would celebrate on a bright Christmas Day.
They all enjoyed freedom, each month and all year,
Because of Marines like this one lying here.

I couldn't help wonder how many lay alone,
On a cold Christmas Eve, in a land far from home.
Just the very thought brought a tear to my eye.
I dropped to my knees and I started to cry.

He must have awoken, for I heard a rough voice,
"Santa, don't cry, this life is my choice.
I fight for freedom, I don't ask for more,
My life is my God, my country, my Corps."

With that he rolled over, drifted off into sleep,
I couldn't control it, I continued to weep.
I watched him for hours, so silent and still,
I noticed he shivered from the cold night's chill.
So I took off my jacket, the one made of red,
And covered this Marine from his toes to his head.

Then I put on his T-shirt of scarlet and gold,
With an Eagle, Globe and Anchor emblazoned so bold.
And although it barely fit me, I began to swell with pride,
And for one shining moment, I was a Marine deep inside.

I didn't want to leave him so quiet in the night,
This guardian of honor, so willing to fight.
But half asleep he rolled over, and in a voice clean and pure,
Said "Carry on, Santa, it's Christmas Day, all secure!"
One look at my watch and I knew he was right,
Merry Christmas my friend, Semper Fi and goodnight!

With all due respect to the author of "Twas the Night Before Christmas," THIS is the poem that should be read to starry-eyed children as they get ready for bed on Christmas Eve, because it lets them know they are free to enjoy the wonders of Christmas Morning and the promise of the American Dream because of the countless sacrifices made by heroes they will never meet.

ARE YOU THERE *Sergeant Major?*

By Captain Jason Grose

"One of the tests of leadership is the ability to recognize a problem before it becomes an emergency." - Arnold H. Glasgow

I share much of Captain Grose's adulation for those who wear the star of a Sergeant Major, although even that lofty rank is infiltrated by the occasional pretender. In my twenty-five years of service there are two such individuals who stand out in my mind, and I must say I was baffled as to how they managed to slip by the promotion board unnoticed. But fortunately they were the exception rather than the rule. When I retired from the Corps I left with a good taste in my mouth, largely because of the last Sergeant Major with whom I served. Kevin Naughton was his name, and when I grow up I want to be just like him! Some of my fondest memories of the Marine Corps are of sitting in Sergeant Major Naughton's office having a cup of coffee and discussing the troops, the Corps and, on occasion, Jane Fonda. He was the prototype, and I hope they haven't broken the mold from which he was made!

I make no secret of my utmost respect for all Sergeants Major. In my view they are not only the top of the enlisted rank structure, they are the very pinnacle of enlisted professionalism. Consider for a moment that of all the thousands of Marines that graduate boot camp every year, only a very precious few will ever rise to the coveted rank of

Sergeant Major and with such competition, the system almost guarantees the Marines who get there are indeed qualified to uphold the reputation of that rank. You cannot fake your way to the top, so when you meet one of these warriors you know you're in the presence of greatness.

There are some that do not hold this opinion, especially officers who see the Sergeant Major as a threat to their authority. Yes, even the most boot Second Lieutenant outranks the most senior Sergeant Major but make no mistake, the Sergeant Major has unspoken power that should never be challenged by anyone. Just try to subordinate the Sergeant Major and you will quickly learn Marines cut from this particular cloth have perfected the art of tearing you down within the bounds of military courtesy afforded to all officers. And if you happen upon one that doesn't even try to stay within the lines, rest assured he has connections to the Old Man who will back him up, resulting in the Sergeant Major telling you where you can stick it. They're at the top with no fear of the promotion race, and they know they really can't be touched. God bless 'em!!

I would like to claim that because of my ten years of enlisted service I knew all of this when I was a brand new butter bar, but I'll have to admit I was just as ignorant as a shave tail fresh out of college. But it was that very enlisted experience that kept me out of trouble with the Sergeants' Major and cemented my relationship with every single one I encountered during my first tour as an officer.

To me, walking amongst Sergeants' Major was to be among gods. When I showed up at Tank Battalion, it was my deep honor to introduce myself to the Battalion Sergeant Major because not too many years before the only time I ever got to even talk to one was if I did something really good - or really bad. Because I never hit either of these extremes very often, my interaction with Sergeants' Major was very limited - which added to their mystique in my eyes.

It was a monumental treat to walk into Sergeant Major Hawkins office and introduce myself, although I found out later it was not a common occurrence since most new officers don't share my adoration for the Sergeant Major. He tentatively stood up to shake my hand, wondering why this old-looking butter bar had come calling. After he found out that I was a prior enlisted Marine he warmed right up, and I felt like the kid who's allowed to sit at the grown-up table for the first time. We hit it off right away, and I was very pleased to find out he had an administrative background. As unbelievable as it seemed to me, I was rubbing elbows with a Sergeant Major who seemed to take me under his wing from the start, and I never hesitated to depend on him for advice.

But I didn't stop there. Almost drunk with excitement, I called up every unit on base and made an appointment with every Sergeant Major aboard 29 Palms. This was quite unorthodox, and most of them were stumped as to what this new lieutenant wanted. When I arrived at their office I would introduce myself and shake their hand while explaining that the reason for the in call was because as a former enlisted Marine, I would never rate to take the Sergeant Major's time like this and I was simply taking advantage of my new officer's status (which would induce a chuckle from both of us). I would then explain to them I was the Adjutant for 1st Tanks and wanted to establish a relationship with the heart of each unit so I could call them up when things needed to get done fast. I may have been a new lieutenant, but I knew that getting the Sergeant Major of each unit on my side would be invaluable during my tour. All of them seemed as surprised as they were complimented by my unique gesture, and it paid off in spades over the next three years. Not only did I get to interact with a collection of my heroes, I was able to get information and assistance faster than any other Adjutant on the base. I was welcomed in every unit, and anything I ever needed was accomplished by a single phone call directly to the Sergeant Major. It turned out my personal joyride to their offices when I first arrived resulted in professional relationships of incalculable benefit.

Not long after I checked into the unit an awful incident occurred which I will never forget. I was sitting at my desk and as often happened, I came across something that I needed the Sergeant Major's advice on. I often made it a habit of running things past him to get the sane man's point of view but on this day, I wasn't sure if the Sergeant Major was in his office or out and about the battalion area as he often was. Without thinking, I barked out "SERGEANT MAJOR!" expecting to hear either a reply or silence. What I didn't expect is what happened.

A few seconds later the Sergeant Major came skidding into my office with a worried look on his face. "Yes, Sir?"

I was completely appalled, and my eyes were the size of dinner plates as the reality of what had just happened set it. I shot out of my chair and immediately started apologizing to him, assuring him I would never, ever, beckon him to my office with such an outburst. I tried to explain to him that I just wanted to know if he was in his office so I could come to him with my question. Because I had never previously yelled out for him (proving I never expected the Sergeant Major to hop to my office on command), he thought something was wrong. He got a kick out of my fluster, but I really was repentant and could think of nothing worse than a Sergeant Major seeing me as someone who thought the metal on my collar outgunned the star on his sleeve - because it doesn't.

FIDELI CERTA MERCES

By Mike Seale

"If you can read this, thank a teacher... If you are reading it in English, thank a Marine." – From a bumper sticker

Last week I sat in on several of the 1st Marine Division interviews which two retired Army colonels now working for RAND Corporation conducted on OIF Lessons Learned. They are writing a history for the Vice Chief of the Army on OIF and they recommended that the document also include Marine Corps and British forces experiences, thus their visit to 1st Division and I MEF.

While at 5th Marines, several of the regimental, battalion, and company commanders involved in the fight in Baghdad recounted some of their experiences. The fight on April 10th for the Amilyah Palace and Hanifah mosque were particularly noteworthy. The 1st Battalion, 5th Marines was tasked with that mission. As the 5th Marines account of the action stated, "Significant enemy action in several locations along the axis of advance and in the objective area, characterized by a relentless barrage of RPGs, along with a torrent of heavy machinegun and small arms fire, resulted in the commitment of the RCT quick reaction force in support of the 1st Battalion. In securing their assigned objectives, 1st Battalion experienced heavy casualties and killed an estimated one hundred Saddam Fedayeen fighters... following 1st Battalion's attack, thousands of Iraqis spontaneously took to the streets of Baghdad to cheer and thank the Marines and Sailors of the RCT for liberating them from Hussein's oppressive regime."

During the debrief to the Division the RAND personnel said that they had no idea that this fight had taken place, the ferocity of it, and the bravery of the Marines until these interviews were conducted. Here are some additional details of the fight that we learned from the 5th Marines officers and SNCOs who had taken part in this engagement. I felt I had to share this with other Marines.

The Battle of the Mosque, as it is known, was actually a nine-hour, intense urban fight. Nearly one thousand RPGs were fired at the Marines and Sailors from windows, doorways, corners of buildings and rooftops. Some of the casualties the battalion suffered were from small arms, and one of the Gunnery Sergeants was killed by small arms through a thin-skinned vehicle. The vast majority of casualties were from RPG fragments. One company reported that their twelve AAVs received thirty-three RPG shots, but that none had caused a catastrophic kill to the AAV. Some of the shaped charge rounds went through both sides of the vehicle, however.

On the first day of the battle the battalion reported thirty-four wounded, most with fragmentation wounds to the head and upper torso. It was only on the day after the battle that the regiment realized the number of wounded was actually *seventy*-four. Many of the Marines had not reported their wounds to their corpsman, because they were afraid that they would be medevaced and not be able to return to their unit in the midst of this intense fight. Illustrating that bravery and devotion to their fellow Marines, a field grade officer in the regiment told us of one young Marine who only went to the Doc on the day after the battle to report severe shrapnel wounds to his left arm, asking the corpsman to look at the wounds and to not say anything because he was losing the use of the limb. The Marine confided to the corpsman that he had been unable to stop the bleeding for the past twenty-four hours. Looking at the blood-soaked dressing, the corpsman asked the Marine how many bandages he had bled through. The answer, "I lost count."

As soon as the regimental leadership found out about Marines hiding their wounds, the word quickly went out ordering everyone who had suffered wounds to have them taken care of.

When I related this story to Irish Egan, he commented, "We still make them like we used to!"

To the 5th Marines, and *all* Marines, Semper Fi!

THESE COLORS DON'T RUN

"It is the soldier, not the reporter who has given us freedom of the press. It is the soldier, not the poet, who has given us freedom of speech. It is the soldier, not the campus organizer, who gives us freedom to demonstrate. It is the soldier who salutes the flag, who serves beneath the flag, and whose coffin is draped by the flag, who allows the protester to burn the flag." - Father Dennis O'Brien, Former Sergeant, USMC

A firestorm over the American Flag broke out recently, and right in the middle of it was, surprise, surprise, a former United States Marine. In a country where the First Amendment protects the rights of those who choose to burn our flag, this patriot is in danger of losing his home simply because he chose to fly it. Is it just me, or is there something wrong with that? What I want to know is, where is the ACLU? They never miss an opportunity to leap to the defense of the poor and the downtrodden. Maybe they were too busy to get involved in this one? Something tells me it would have been a different story if this flag owner had struck a match, instead of install a flagpole. The following story, entitled "Community Rallies Around Jupiter Flag Owner," appeared on July 13 of 2001:

Veterans and other citizens rallied around a man in Jupiter, Florida who was being asked to take down the American flag from his flagpole. George Andres doesn't want his flag to touch the ground, and that's why he installed the flagpole. But the homeowner's association said that the pole wasn't appropriate in their community, a judge agreed with the association, and Andres now has to pay a one hundred dollar fine for each day the flag flies.

Supporters from across the country have helped Andres cover the daily fine, but he still owes more than four thousand dollars. A local radio station has planned to set up camp all day at his home, until enough money is raised to cover his fines. Andres has another court date one week from Friday.

But that wasn't the end of that. A second story, entitled "Marine Vet Could Lose Home for Flying Flag," appeared two years later on September 11, 2003:

"A Marine veteran who violated neighborhood association rules by flying an American flag in his yard could lose his home next month. A judge said that the neighborhood association could sell the home in order to collect legal fees against the vet. George Andres vowed to appeal the ruling in a legal saga that has spanned three years and drawn support from Florida Governor Jeb Bush and State Attorney General Charlie Crist.

It is the second time in recent months that Andres' home has moved toward the auction block. He was granted a reprieve in May when Circuit Judge Edward Fine agreed to reconsider his order authorizing the foreclosure. Crist's office argued that Andres' home was constitutionally protected from foreclosure by a homeowner's association under the state's homestead law. But Fine rejected that argument.

Another judge ruled that Andres didn't have a right to put up the flagpole. The association filed a lien on the property to collect the roughly twenty-one thousand dollars in attorneys' fees and legal expenses it spent to win the case."

That such a battle would be waged by that man's neighbors is simply appalling. Perhaps they would also like to amend the Marines' Hymn to say "Our flag's unfurled to every breeze, from dawn to setting sun (except in Jupiter, Florida)." Finally, I would like to refer the judge, attorney, and Homeowners Association to the quote at the beginning of this story. Shame on them!

OOH-RAH!

By Vince Crawley

"An army of sheep led by a lion are more to be feared than an army of lions led by a sheep." - Chabrias, 370 B.C.

They say imitation is the sincerest form of flattery, and while I suppose that is true there are times when people and organizations just need to remain in their own element.

I made it quite clear what I think of the Navy's Special Warfare program in 'Swift, Silent and Surrounded.' It is literally a case of "fish out of water." The Navy needs to stick to ships and submarines, and leave land warfare to the experts.

Now the Air Force is trying to act more like the other services, and it isn't necessarily all bad. For one thing, they are finally implementing a Physical Fitness Test that involves running instead of riding a stationary bicycle. I guess someone finally realized they wouldn't be riding Schwinns into combat! But they have made a few well meaning, but ridiculous, decisions as well.

One is the new Air Force camouflage uniform. When I first saw it I thought someone was playing a joke on me. It is a tiger stripe pattern which includes, among others, the colors blue and gray. Exactly what do they expect to blend in with - clouds?

And now the Air Force is trying to manufacture a "war cry." How pathetic. What they really need is for a few Gunnery Sergeant Hartman-like Drill Instructors right out of Full Metal Jacket to infiltrate their basic training, get in the face of a few airman recruits, and scream "Let me see your war face!"

It may sound like so much "hooah" to some, but a lot of folks in uniform take their guttural war cries quite seriously.

So ground-pounders all over the planet are up in arms about an Air Force suggestion that airmen should be shouting "airpower!" in place of the more earthy hooah!

The phonetically spelled battle cry "hooah" - or its Marine Corps equivalent, "ooh-rah" - often is barked when troops want to voice approval or a sense of esprit de corps. Its full meaning is primal and difficult to define, for it also echoes the hardships faced by those in uniform.

Soldiers tend to prefer hooah. The Marines insist there is a separate and distinct ooh-rah. Not only that, they claim theirs was first. While the Army can trace hooah back only to the Second Seminole War of 1835, the Marines cite Revolutionary War battle cries and even Russian and Turkish precedents for ooh-rah, which holds tremendous meaning and significance for most Marines.

Just listen to Gunnery Sergeant Glenn Holloway, a combat correspondent based at the Navy Annex in Arlington, Virginia:

"Ooh-rah comes from the places in our hearts that only Marines understand. It is conceived in sweat, nurtured with drill. It is raw determination and gut-wrenching courage in the face of adversity. It is a concern for fellow Marines embodied by selfless acts of heroism. It cannot be administrated. It is not planned and put into action. It cannot be manufactured. Ooh-rah must be purchased. Ooh-rah is Marine."

The Navy, generally satisfied with its own time-proven "aye, aye, sir!" – which reaches back to Elizabethan times - remains on the sidelines of this debate.

Air Force Colonel Jay DeFrank, director of Pentagon press operations, said he's unaware of a top-level push to promote airpower over hooah in the Air Force. But he said he has heard a lot of airpowers bandied about lately, usually in conjunction with Air Force gun-camera footage taken over Afghanistan and northern Iraq.

Researching the matter, he learned that retired Air Force General Lloyd "Fig" Newton pitched the airpower utterance during his stint at Air Education and Training Command in the late 1990s. But DeFrank said he was told Newton's idea had never quite caught on - until now.

In November, the idea of adopting airpower as the service's battle cry was presented to Air Force Chief of Staff General John Jumper by a group of security forces airmen, according to Air Force spokesman Lieutenant Colonel Tyrone "Woody" Woodyard.

Jumper "clearly is an advocate of air power," but he has no preference when it comes to his airmen shouting airpower or hooah, Woodyard said. "General Jumper supports anything that unifies, inspires and motivates a unit to complete its mission."

Woodyard did add that the chief does not go around shouting hooah! himself.

E-mails bouncing between the Air Force and Army special-operations communities shed more light on the unfolding debate.

A message out of an Air Force special-operations command in the Persian Gulf region in September lays it out: "By now, most of you have heard that the term hooah! is not encouraged in our Air Force. If you are looking for something to say in those times of great excitement and agreement when hooah! seemed to fit right in, try a good solid airpower! Airpower will always be uniquely Air Force."

This led to predictable Bronx cheers from the rank and file. "Why would a simple word that means so much to so many take up the time of people who have so much more to worry about?" asked one seasoned Air Force member who signed his message: "HOOAH!!"

"Hooah simply means everything except 'no,'" said another Air Force advocate of that word.

Another message, from Ralph Hatcher at Dyess Air Force Base, Texas, offers further explanation about hooah, which Hatcher said is used "by all Special Forces types… however, can also include anyone that performs their duty knowing lives depend on what they do."

In an interview, Hatcher said that while deployed to Hungary in late 1997, he bought a T-shirt that had fourteen definitions for hooah. Among them:

Good copy, solid copy, roger, good or great; message received, understood. I do not know but will check on it. I haven't the vaguest idea. I am not listening. That is enough of your drivel. Sit down. Stop sniveling. I don't know what that means but I am too embarrassed to ask for clarification. Amen.

Army Private Ramon Gomez said he approves of the Air Force's airpower slogan.

"I think it's good because no one really knows about the Air Force, just the airplanes. That'll give them some distinction," said Gomez, a member of 1st Battalion of the 325th Airborne Infantry Regiment at Fort Bragg, North Carolina.

Army Sergeant Todd Wilson has a different opinion.

"That's weak," said Wilson, a senior instructor at the basic noncommissioned officers course at Fort Benning, Georgia, who likened airpower to something that might come from the Powerpuff Girls, heroines of a popular TV cartoon.

These days at Fort Benning, spiritual home of the Army's infantry, enlisted, officers and even civilians hooah one another at meetings, in the hallway or during training.

"I even heard a Marine say hooah once," Wilson said.

Marines, however, would beg to differ. Their ooh-rah, they claim, is uniquely their own and exists as a separate and distinct word.

Navy people are uninvolved in all this hooah-upsmanship.

"Nah, we don't say hooah," one Navy officer explained, barely uttering the word above a thin whisper.

Another officer, Navy Lieutenant Dan Hetlage, said the problem has come up in joint-training environments when sailors want a battle cry of their own. While based at the Defense Information School, Hetlage said, he and his fellow sailors experimented with pirate-like sounds.

"'Aaarrrrhh' was the best we could come up with," he said. But he readily acknowledges that this didn't really convey the correct meaning.

This story originally appeared in the Air Force Times in 2002.

PRISONERS
Of Political Correctness

By Jane Chastain

For those who may not remember the story, Jessica Lynch is a former Army Quartermaster Corps Private First Class who was taken prisoner during the 2003 invasion of Iraq. She was rescued by United States forces on April 1, 2003 in what was the first successful rescue of an American POW since World War II and the first ever of a woman. Lynch, then a nineteen-year-old supply clerk with the 507th Maintenance Company, was injured and captured by Iraqi forces after her group made a wrong turn and was subsequently ambushed on March 23, 2003 near Nasiriyah,

Thank God Army PFC Jessica Lynch has been rescued from an Iraqi hospital where she was being held as a prisoner of war!

As everyone knows, Lynch was part of the 507th Ordnance Maintenance Company that fell into an Iraqi ambush, which led to the deaths of some of her fellow soldiers and the capture of a least five others, including Specialist Shoshana Johnson. The fate of PFC Lori Piestewa and several others still is unknown. If your heart wasn't in your throat when you saw the pictures of Jessica on a stretcher, or a wounded Shoshana being interviewed by her captors, then it's time for a reality check.

Something is terribly wrong when the most powerful country on earth is assigning women service members to units where they are subject to capture, rape, torture and death, while able-bodied men are stationed out of harm's way or, worse still, at home in the comfort of their living rooms. Guys, do you hide under the covers and send you wives downstairs if you suspect a burglar is in your home?

We look down our noses in disgust at Saddam Hussein's disregard for human life - and his brutal treatment of women - but we are deliberately sending our young women into combat zones so that they can be sacrificed on the altar of political correctness. Time for a reality check! The feminists celebrated the news that Johnson had been taken prisoner and put on public display. Alas, another trophy on their road to prove that men and women are interchangeable fungibles! "The capture of this woman," they croon, "proves women are just as brave, capable and well-trained as men and have just as much chance to survive." That, of course, is rubbish!

It may not be *fair* that a man is, on average, six inches taller, thirty pounds heavier and - more importantly - has forty-two percent more upper body strength, but it is reality. The dirty little secret in the service academies and our boot camps is that women are passed right along with the men because of "gender norming" - where the emphasis is on "equal effort," not equal results. While the numbers are fudged to make everything come out equal in these controlled environments, these same women will not have an equal chance to survive on the battlefield. That is why women are not supposed to be assigned to ground combat units.

So, how is it that Lynch and Johnson - who were trained as a file clerk and cook, respectively - were assigned to a unit that was ordered into the heart of Iraq? A lot of the blame can be laid at the feet of our serial philandering former president, Bill Clinton, and his secretary of defense, Les Aspin. In 1994 Aspin redefined direct ground combat by eliminating "inherent risk of capture" as a factor in deciding whether a unit was judged to be "close combat" or merely "combat support" in order to open up more "career opportunities" for women.

This was a cold, calculated political decision. Enlisted women like Lynch, Johnson and Piestewa were considered expendable in order to serve the needs of women officers, who would use their deaths and capture as stepping stones on their way to the Joint Chiefs of Staff. The important thing to remember is this: There was no shortage of opportunities for women to serve in the military then - and there is no shortage of men who can serve in battle zones today. This is not about giving Army women the choice of whether they want to be assigned to units in battle zones. Soldiers cannot pick and choose their assignments. If women *can* be assigned to these units, they *must* be assigned to these units.

However, it is Congress that makes the laws governing our military. Therefore, the blame must be laid squarely at the feet of these lawmakers, both Democrat and Republican, who find it a lot easier to sacrifice enlisted women than undo the damage and have to face the ire of a handful of radical feminist lawmakers they see every day on Capitol Hill.

It is time our lawmakers forget about political correctness and face the realities of keeping the men - who must do the heavy lifting in these units - alive, and keep the women, who are providing invaluable support services, out of harm's way to the greatest degree possible. To do anything less is pure cowardice.

TELL IT TO THE MARINES

By Beth Fallon

"The legitimate object of war is a more perfect peace."
- William Tecumseh Sherman

The twentieth anniversary of the Beirut bombing is coming up next month, and as I prepare to travel north to Camp Lejeune and the Beirut Memorial I can't help but think about the world as it was on that dark day in 1983, and the state of the Middle East both then and now. I am sad to say not much progress has been made, and in fact things may actually be worse. It is difficult to remember those peacekeepers who died on October 23rd, 1983, and then turn on the television and see even more of the same hatred and violence they gave their lives trying to prevent. Did they die in vain? I don't know. But in light of the events of the past couple of years I find the following article, written in the wake of that bombing, somewhat chilling - because the more things change, the more they stay the same. Yasser Arafat. Failed diplomacy. Suicide bombings. The beat goes on:

"Tell it to the Marines," they used to say. They have been saying it, in fact, since Charles II of England (1660-1685) remarked that, since his Marines had been everywhere and seen everything, he would check out any unbelievable story he heard with them. "Tell it to the Marines," people said, as the meaning evolved. Take that piece of bull you're peddling, and shove it. Tell it to the Marines.

There are people running around this week saying what a terrible thing it is for the United States of America to fight with guns in its own interest. "An act of war," condemned

Senator Daniel Patrick Moynihan (D-NY) of the US-Eastern Caribbean invasion of Marxist Grenada. There, one of the Marxist factions had just amused itself by publicly exterminating the Prime Minister, three Cabinet members and two labor leaders from a rival Marxist faction.

Tell it to the Marines, Senator. Tell it to Jean Joel of Albany, NY who responded to what apparently was hostile skepticism from reporters in South Carolina when the first of the evacuated students landed. Were they really in danger? Did they really feel threatened?

"I don't know what you want me to say," Jean Joel replied, her eyes narrowing. "The only thing we were sure of was who was helping us - and that was the Marines who saved us."

"I have been a dove all my life," chimed in Jeff Geller of Woodbridge, NY. "And I don't want to hear anyone say one word against those Marines. Those Marines came down and saved us." Network correspondents, irked at being kept on Barbados, went looking for local comment. In Barbados and Jamaica, the invasion was "welcome," said Sandy Gilmore of CBS. "We're free, mon, free," exulted a Grenadian. Unpopular? Not in *that* neighborhood. Tell it to the Marines.

Tell it to the Marines in this most poignant of weeks. Tell them while they were pulling 221 bodies out of their terrorist-bombed headquarters in Beirut, why United States Marines were ordered to patrol a war zone with empty guns. Tell it to Lance Corporal Robert Calhoun, who got blown off the roof in the Sunday bombing, and who heard this account of the death truck from a surviving Marine sentry.

"The man was wearing green fatigues," Calhoun said the sentry told him, "and as he went by him, he tried to pull out a magazine because it wasn't in his weapon, but by the time he got everything loaded the man had already exploded the weapon. And there was nothing he could do. And he said just as the man went by, he says he'll always remember, the guy was smiling."

Have you read that already this week? Good. Read it again. And again. Read what happens to people when leaders fantasize that you can fight terror with virtue or aggression with good will. Read what happens when politics becomes make believe. Read what happens when you try to keep the peace with empty guns.

This is a tough week, and not a time to lose your temper. I'm not saying that you should always fight, or fight for everything. You've got to choose. I would not spend one American life, much less 221, plus six who died before, plus the total of fifty-four French dead, for the murderous warlords of Lebanon. One and all, they disgrace humanity by joyously killing anybody who doesn't call God by the same name they do, with particular and special reference to the Israelis.

Basically, we are in Lebanon to secure the safety of an ally whose security we guarantee, to pull Israeli irons out of the fire. So where are they? It is they who have the most to lose if Syria succeeds in dominating a partitioned Lebanon, and hands a nice new base to its Soviet allies. The Muslim world hates Israel, and the Soviets exploit that. But we have plenty to lose too, in a Middle East threatened with Soviet domination. So we can't afford the fantasies and waste that led to the debacle in Vietnam. That sorry slaughter proved that if you can get a livable deal at the table, fine. If you can't wars have to be fought all out to be won.

I don't like Marines dying in any event. But I sure don't want them dying for nothing, or doing any surrogate dying. "Semper Fi," one shattered Marine scrawled on a hospital pad for Commandant P. X. Kelley. "Always Faithful."

What an indictment. They will sit there like ducks and die and not complain. But I'm complaining, as loud as I can. Decide what you need and can get, based on reality. Then go in and get it, whatever it costs. Or get out. And don't tell me that you can keep the peace with empty guns. You can tell that to the Marines - if you dare.

This article originally appeared in the Navy Times in 1983.

CLOTHES MAKE THE MAN

"Every girl's crazy 'bout a sharp dressed man."
– From the song by ZZ Top

I have a lot of respect for fighter pilots. I really do. Anyone who provides close air support for me, and for the rest of my ground-pounding buddies, is okay in my book (no pun intended). But I do take a bit of exception to the air of superiority some of them like to project. I realize it takes a lot of training to fly a jet aircraft, and a great deal of skill to land on the deck of an aircraft carrier in the middle of the night, but let's face facts. When their missions are over the fly-guys head back to the O-Club and swill Scotch, while the mud Marines are still out in the boonies humping rucks and fighting hand to hand. The pilots like to point out that is the case because they are smarter than we are. That may be true. But it should also be pointed out that while there are a growing number of female pilots, there are no women in recon or the infantry. That always seems to take them down a couple of notches.

During the time I was a student at Scuba School in Little Creek Virginia I was witness to a strange phenomenon for the first time. A couple of the officers in my class invited me to accompany them to happy hour at the Oceana NAS Officer's Club on a Friday night as their guest. I could see immediately why they wanted to go there. It was a meat market! After one look at all of the women who had traveled long distances to be there, and who were dressed to kill, I couldn't help thinking about the movie *An Officer and a Gentleman.* You know, the part where the blonde bimbo says, "I don't want to marry no Okie from Muskogee. I want to be the wife of a Naval a-vi-a-tor!" But it was in the parking lot, prior to reaching the front door, that I was treated to an amusing spectacle. Dozens of men, presumably pilots, were struggling to pull on flight suits over whatever it was they had been wearing up to that point. It was comical. I didn't really understand what was going on at the time, but soon figured it out. Every one of those guys was trying to give the impression he had just come straight from the flight line after an arduous day in the cockpit. Geez. Now, it's no secret that Marine Dress Blues are a "chick magnet" – in fact it is almost impossible to look bad while wearing them. But I have never pulled on a set in a parking lot on a Friday night. I guess those guys just don't have any game unless they are ensconced in Nomex.

Once our group got inside we decided the best policy was "when in Rome, do as the Romans do." We each began to using our hands to demonstrate dog fighting tactics while conversing with each other - it made no difference what we were talking about, as long as it *looked* like we were talking about flying. Those who had them also pinned miniature sets of gold jump wings on their lapels, and I was surprised how often the civilians present mistook them for pilot's wings. At one point I even had one bleached blonde convinced they signified my position as a door gunner on the space shuttle. Hmm… I think I'll just stick with Dress Blues.

BIT OF COLORED RIBBON

"Do not attack the 1st Marine Division. They fight like devils. Strike the American Army." - Chinese Army directive during the Korean War

Are soldiers and airmen braver than Marines? If the fruit salad those guys are wearing on their chests is any indication, they must be! The Army has long given out all kinds of ribbons and badges for throwing grenades, running the bayonet course, attending NCO school, and getting a jeep license. They even have a ribbon (along with the Air Force) for graduating basic training, and not long ago decided ALL soldiers rate to wear the once-coveted black beret. While that is for the most part laughable to Marines, the way personal decorations are handed out by our sister services can only be described as tragic.

In the years between Vietnam and the Gulf Wars it was not unusual, in fact it was commonplace, to encounter Marines with only one or two ribbons on their uniforms. In the Corps our "badges" are limited to those showing a Marine's level of weapons qualification, which is a cornerstone of our ethos (every Marine a rifleman), and to a few associated with hazardous duties such as pilots, parachutists, divers and explosive ordnance disposal technicians. I believe all of them are appropriate.

Personal decorations are a different story altogether. Personal awards, particularly those for valor, must endure intense scrutiny in order to make sure they are appropriate – and that is as it should be. It is unfortunate that an undeserved decoration occasionally slips through the cracks, but no system is infallible. Is the Marine Corps immune from "awards abuse?" Certainly not. But when you see a Marine wearing an award the odds are pretty darn good he earned it.

The Army, on the other hand, believes heroes can be manufactured, and that awards are a stepping stone to promotion. One of the best examples was the invasion of Grenada in 1983. It was a minor skirmish compared to most conflicts, and in fact the U.S. Army did not acquit itself very well at all. Even so, more than 8600 awards were given out despite the fact there were only about 7000 soldiers on the island! The Marine Corps, on the other hand, awarded only a handful despite kicking ass and taking names.

They just don't get it. The Army doesn't understand when you give something away cheaply it loses its value, as well as its meaning. You will not have to look any farther than Jessica Lynch being awarded the Bronze Star to know that is true. I am beginning to believe it is harder to get a Marine Corps Good Conduct Medal than an Army Bronze Star.

Consider the following, which was written by retired Army Colonel David H. Hackworth. Keep in mind he was the most highly decorated living American at the time he penned it, and therefore uniquely qualified to weigh in on this subject. His observations:

Recently in Iraq, an Army two-star general put himself in for the Silver Star, a gallantry award, for just being there, and for the Combat Infantryman Badge, an award designed for infantry grunts far below the rank of this division commander. During the war, members of an Air Force bomber crew were all awarded the Distinguished Flying Cross for lobbing a smart bomb from 30,000 feet onto a house where Saddam was rumored to be breaking bread - even though Saddam was still out there somewhere sucking desert air. In 1944, the only way a bomber crew might have gotten the DFC would have been if it had wobbled back from

Berlin on one wing and a prayer after a dozen-plus missions facing wall-to-wall flak.

Here's another "Believe It or Not." When the Scuds were thumping down on Kuwait, a Navy two-star admiral and six of his flunkies were awarded the Bronze Star after a missile struck ten miles away.

Not that these abuses of the awards system are anything new. The U.S. military's awards program - designed to recognize both our combat heroes' valor and the meritorious deeds by those hard-working supporters who bring up the rear - has never been exactly fair. In the past, Joe and Jill have often gone unrecognized because there was no one left at the end of the battle to bear witness, or the paperwork got lost or wasn't written persuasively enough, or some eager-beaver officer in the chain of command stole their glory.

I know of two Medals of Honor - one in Korea and the other with a Navy unit in Vietnam - that shamefully went to still-living former officers when in fact their above-and-beyond deeds "witnessed" by sycophants were actually performed by grunts.

In the latter days of the Korean War and in Vietnam, Grenada, Panama, Desert Storm and Somalia, such abuses of military honors increased with each battle. In Vietnam, a dog was awarded the Bronze Star, and in Grenada, more medals were awarded than there were soldiers on that tiny island. In Desert Storm, Army infantry battalions that never saw a shot fired were awarded the coveted CIB. Now warriors in Iraq are reporting that COs there are using a quota system for awards.

Sergeant Bill Casey, whose unit saw heavy combat in Iraq, says: "Our awards were not given out for heroism. They were based entirely upon rank and duty position. If you were a company commander, you got a Silver Star. If you were a platoon leader or platoon sergeant, you got a Bronze Star. If you did a good job at a level below that, you might get a Bronze Star. If you were a PFC, you probably didn't get a medal for valor. Every award was entirely based upon rank and duty position - rather than whether that person stood tall and continued to return fire or whether that person continued to bring the fight to the enemy or flat-out ran away when the bullets started flying."

These stats tell the story: The U.S. Air Force has approved more than fifty thousand medals for operations in the Middle East. The U.S. Army, trying to catch up with the folks in blue who flew through all that imaginary Iraqi flak, has issued medals as though they were Cracker Jack prizes. So far they've pinned on tens of thousands of awards, from the coveted Distinguished Service Cross to the CIB. More than five thousand Bronze Stars alone have been awarded. One-half the members of a 700-strong aviation squadron at Fort Stewart, Georgia, were recently presented Bronze Stars and Commendation Medals.

But as of September 22, 2003 the Marine Corps had approved only fifty-six Meritorious Bronze Stars – forty-six to officers, ten to enlisted - and fifteen Bronze Stars for valor - eleven to officers and four to enlisted - for their seventy thousand fighters who kicked more than a little butt during the war in Iraq.

Kudos to our gallant Marine Corps for not following the quota system, and to its top brass for refusing to play the Pentagon's public-relations medal-giveaway game.

But any way you count 'em, deserving grunts aren't being appropriately recognized by a sick, out-of-control system that desperately needs overhauling.

The following back-channel e-mail was written by a Marine Corps General Officer after he read Colonel Hackworth's article:

Have received this today from several folks, and it's something that has been bantered about back-channel for the past several weeks. Please indulge me with regard to the following commentary:

It's very reassuring to see in print (I MEAN HECK... we're talkin' Hack's column here so....) that our Corps continues to hold the line on awards stemming from actions in this latest series of actions. Over the past few weeks I have spoken to several General Officers of Marines about this very same issue and "our" concerns... some just recently here in the Camp Lejeune area, and some at the very highest levels of our Corps. They too say that the awards for this show are being scrubbed quite closely... as they should be.

One of our young Generals, who was amid much of the action in Iraq, told me about a young enlisted Marine who was maneuvering his AAV within the moving combat formation... he saw a bad guy pop up with an RPG... and instead of maneuvering away from the danger, he did as we'd expect of most young Devil Dogs... he put the pedal to the metal and took that sucker out by running right on over him! His Navy Commendation Medal was just recently approved... maybe even awarded by now. Some might say he was just doing his job and, of course, they'd be right... but the WAY he went about doing his job makes all the difference in the world...!

A concern some folks have brought to my attention, and again something that I have shared with our most senior leadership, is that "some folks" are saying (no... I don't have names and it could just be a rumor) that the Combat Action Ribbon may end up being given out on the cheap.

While I hope that's NOT the case, some of us who served in RVN, for example, can well-remember how that went... ya know... someone heard the sound of an incoming/exploding arty/mortar/rocket round (or maybe just knew someone who had heard of same) and they end up having the CAR appear on their little pigeon chest as well as on their page 9...! Some of you out there know what WE did to come to "rate" the CAR (as in "look 'em in the eyes whilst putting well-aimed small arms on their scrawny little bods"), and while there are varying levels of actual combat actions involving the enemy that can have one become qualified for the CAR, merely hearing stories about combat ain't one of 'em...! Let's just make sure we don't let it get outta hand this time around.

Medals, medals, and more medals... we got folks over here with Medals of Honor, Navy & Distinguished Service Crosses, Silver Stars, Bronze Stars, various Commendation and Achievement medals, Purple Hearts... all awarded from their actions in combat... Hack's story is just really gonna set some of 'em off just as it did me...! Semper Fi.

TANGO YANKEE

By Christopher S. Dowling

"We must indeed all hang together, or most assuredly we shall all hang separately." – Benjamin Franklin

In Pat Conroy's novel 'The Great Santini' he made an observation that the only times the Navy and the Marine Corps are on the same side is in time of war, and during the Army-Navy game. A slight exaggeration to be sure, but really not that far from the truth. After all, we love to characterize Navy CPOs as being overweight, donut eating, coffee drinking slobs - whether it is true or not. We also wonder aloud if the inclusion of a sailor in the gay singing group The Village People was an accident. Marines generally don't think so. And those uniforms they wear! How could any heterosexual man possibly wear cracker jacks? After all, who can forget the scene from 'A Few Good Men' where the Marine Colonel portrayed by Jack Nicholson tells Navy officer Tom Cruise to "stand there in that faggoty white uniform and show me some respect!"?

If all of that sounds harsh, keep in mind our salty friends in the Navy like to have a bit of fun at our expense as well. But all of those inter-service fun and games come to a screeching halt the moment the first round goes downrange. The Navy-Marine Corps team is unmatched when it comes to projecting power ashore, and both services adhere to the traditions of the sea. Most importantly, anyone who disrespects one of our beloved Navy Corpsmen in the presence of a Marine is signing his own death warrant!

The letter below, which was written by a Marine Major, is a recent example of that bond between the Sea Services. No doubt there was a bit of good natured needling going on just moments before the attack in question, but what occurred afterwards speaks for itself:

Yesterday afternoon around 1510, some of you may have seen me standing in front of my office with a female Navy Petty Officer 1st Class. She was wearing her dungaree uniform. She was shaking, she was crying, and it was obvious that she was in severe emotional pain. You may have seen me hug her, you may have seen us talk for about four minutes until she turned and left the building. Four minutes is not very long, but those were four of the most eye-opening minutes I have ever experienced as a U.S. Marine.

The Petty Officer entered the front hatch of MATSG-33 looking confused and distraught. Thinking she was just another sailor looking for directions somewhere aboard NAS Oceana, I walked out of my office and greeted her and asked if I could help her. The name on her shirt said "Stewart." PO1 Stewart remained silent and stationary, staring blankly at the deck. I asked her if everything was okay. Her hands started shaking and her bottom lip started to quiver as tears began streaming down her face. She just stood there, clutching her cover tightly in both hands as she cried silently for about twenty seconds before she could manage to get a word out.

I was feeling helpless at this point because I had no idea what to say to her without knowing what was wrong. After she told me, I still had no idea what to say. I was just proud to be a Marine.

Through choked-back tears, PO1 Stewart told me why she came to MATSG-33. She said she was talking with four of her closest friends one day while they were on ship last October. Their ship was the *USS Cole*.

She said that it all happened so quickly. One moment they were talking as usual and the next moment, all four of her friends were lying beside her, and she was the only one alive. PO1 Stewart said the real terror sunk in moments after the explosion, after she saw the dead, soot covered bodies of her friends, when she realized that at any moment, another explosion may take the lives of more of her shipmates or her own. She said she was so afraid that the terrorists weren't finished with them yet. Then she saw the Marines. The Marines came and secured the area. The Marines came and secured the survivors. PO1 Stewart said that she knew, and everyone aboard *USS Cole* knew, that the terrorists had got their one deadly shot in, but no more lives would be lost that day while the Marines were there.

I know that it was one of the FAST companies that responded that day. PO1 Stewart only knows that it was the Marines. I used to be an infantryman and part of the Marine Security Force, but that was five years ago. I have never set foot on the *USS Cole* or patrolled its surrounding waters. The day *USS Cole* was bombed, I was sitting at a desk doing paperwork on a quiet Navy Base in Virginia Beach. Yet on an ordinary summer day, a Navy Petty Officer 1st Class who felt the explosion of the *USS Cole* and saw her shipmates die before her, walked into Marine Aviation Training Support Group-33 to find any Marine who she could look in the face to say thank you.

I was choked up and absolutely stunned by what I had just heard. I hugged PO1 Stewart and offered to contact the FAST companies to locate the Marines who responded that day, but she told me that she was retiring this week and this was closure for her. By saying thank you to a Marine, she is ready to try and move on from her nightmare. I told her that I would extend her thanks. PO1 Stewart said thank you once more, turned and walked out of MATSG-33. I sat back down in the chair of my quiet office and continued my paperwork - with a much better view of the big picture.

From PO1 Stewart, formerly of the *USS Cole* – "Thank You, Marines!"

Note: TANGO YANKEE is the Semaphore version of "Thank You."

POSTCARD FROM THE EDGE

"Those who expect to reap the blessings of liberty must, like men, undergo the fatigue of supporting it."
– Thomas Paine

Civilians seem to complain about everything from breaking a fingernail to getting a scratch on their Mercedes. Life is SO hard for them. I for one have learned to appreciate the simple things in life like hot water, clean sheets and a real hamburger, and pity all of those spoiled souls who just don't know when they have it good. Marines do things on a daily basis that would be inconceivable to a yuppie, and for us it's just another day at the office. Even after all my years in the Corps it never ceases to amaze me how adaptable Marines are, and how they seem to thrive in the worst of environments and situations. Consider this e-mail home from an officer serving at Camp Babylon in Iraq:

Well, it is the heart of summer and the middle of our overtime period here in Iraq. We hear that summer finally arrived there in California after a full game of "June gloom" and that you folks in Arizona are having quite a humid period as well. Weather-wise here, there is no change from hot and more hot, but we have been all laughing at our likely condition when we get back. We haven't seen temperatures below a hundred in the day and eighty at night since early June, and think we will be racing for sweatshirts when we get back home and it is a perfect seventy-two degrees outside. We look forward, however, to re-acclimating to the real world.

We continue to send nearly everyone on a mission on most every day up here in Iraq. Even the Marines who thought they were going to coast easy on my Brigade Platoon staff have been going out regularly to accomplish our Brigade Platoon goals. We are down to only four Marines full time in Kuwait holding the fort at Camp Commando, two of them working for the MEF (Marine Expeditionary Force) Headquarters Group there. The rest are on Operational teams or keeping us functioning up here at Camp Babylon.

And now, a word from our sponsor. I'm not exactly sure what the media is saying back home, but your Marines and Corpsmen are out here doing what needs to be done, just like they were while it was still a war. Regardless of how the WMD question washes out, the Iraqi people were extremely oppressed by the former government, and this country has been falling apart for the last twenty years. This is a bad neighborhood in the global village and the people show no gratitude now, but it is a very worthy cause to set them free and get them to start taking care of themselves. As it is now, they have been tied to the government for EVERYTHING, and show no ability to take care of themselves. Of course, it will be a lot nicer when the Multi-National Divisions take over from us and we are home reading about it in the newspapers next to you.

Mostly now we are hired-gun teams running escort for different MEF taskings. We escort important people, equipment, and supplies within the Marine Corps area of operations. Sometimes we are just a small three-vehicle convoy, two of our teams and the vehicle we are escorting, and sometimes we escort over fifty vehicles along the barren Iraqi highways. It is a lot of long, hot driving in evil winds, and one of the essential pieces of equipment the teams must grab before departing for their full day on the road is an ice chest or two full of water, soda, juice… and, of course, ice. Have I mentioned an interest in ice to you before? We love ice, and openly fantasize about being in a place where we have clean ice to put INSIDE the drinks and not just pack AROUND our bottles and cans. As you might imagine with a bunch of warriors, we also talk a lot about drinks that don't require ice in the glass, and come out of a cold TAP!

Besides the ice, you may have noticed a certain obsession with air-conditioning. In this heat, air-conditioning changes everything about the way we see the world. We kneel down before the great AC idol and have sacrificed ALL the women in the platoon to that deity. Apparently, our lack of suitable sacrifice has not gone unnoticed by the Air Conditioning idol, and we have yet to go twenty-four hours without a breakdown, shutdown or meltdown. Our idol has a sense of humor, and shuts the system down regularly between noon and 1600 for at least an hour or two when the day's heat has peaked. Two days ago we received a new fancy civilian generator almost twice as big as the old one. Bigger must be better, but just to make sure we know who's boss, that generator went down after running for about twelve hours, and NEVER ran during the heat of midday. Now we are using an old Marine green unit that has run for about half an hour all day. They keep coming over to brief me that "it's up for good this time." Yeah, really good sense of humor!

Even our equipment knows it is time to be home now. The vehicles and radios are starting to resist their overuse these days, and we would be lost without our Motor Transport and our Communications Marines. Regardless of what they did before this phase of the deployment, those Marines have all become technicians and repair gurus this time around. Of course, we have also discovered what their families probably learned a long time ago: a lot of these Marines are happiest with some broken piece of equipment before them to be torn apart, fiddled with, cleaned, tweaked, repaired, sharpened, adjusted, strengthened, tightened, and finally rebuilt. Tim Allen the Tool Guy has NOTHING on our Marines.

Okay, what rumors are you all getting back there? Well, I wish I had better information to give you, but I get the brief straight from the general, and I'll tell you what… I STILL DON'T KNOW EXACTLY WHEN WE ARE GETTING OUT OF HERE! In the last two days we went from getting down to Kuwait a few days early, around August 24th, to getting down there a few days late, around September 4th. Even the straight scoop delivered in fancy-schmancy, full-color, multi-slide Power Point Presentations still means nothing two days after we get the brief. There are a lot of moving parts those big guys have to account for, so we will see what finally solidifies, but for now, nothing solid to report. Go ahead and just believe your rumors if they are good, as they are probably as accurate as anything right now, in fact, send us your really good rumors and we will try to pass them off as "ground truth" from this end. I will keep you up-to-date, but for now, stick with that second week in September for when we'll get home. "Home for Labor Day" is still a nice fantasy though.

Thanks to all of you who sent care packages to us out here. We go quickly through those goodies and baby wipes, and love those great powdered drink mixes. We still miss you, miss you more, and everything about home. Thanks for the great support!!

Semper Fidelis,
Chewy
LtCol M.C. "Chewy" Vacca
3rd ANGLICO, 4th BdePlt

ANIMAL HOUSE REVISITED

"The only easy day… was yesterday!" – USMC Axiom

One of the unique things about the MSG Program is the way the Marines are billeted. Naturally the Detachment Commander is provided with his own house, since only Staff NCOs are allowed to be married while on the program. The rest of the detachment must agree to remain single, and they live together in a facility known as the Marine House. For 1/5 Dets (one SNCO and five watchstanders) that usually ended up being nothing more than a big house, but for the larger detachments the Marine House could sometimes resemble a small college dormitory.

One of the big challenges for a Detachment Commander is enforcing the understandably unpopular battalion regulation which prohibits members of the opposite sex (in most cases that means women) from entering the rooms of individual Marines. It is quite a challenge because it is essentially a battle between regulations on a piece of paper and the raging hormones of a twenty-something Marine, and Mother Nature can be a pretty formidable adversary.

For the most part that rule wasn't a big problem for my detachment in the Congo, and I am willing to bet the same was true for Marines assigned to many other third world posts. Aside from the occasional airline stewardess in transit, there were virtually no women to be found in Brazzaville - at least none who spoke English, who bathed regularly, or who had not contracted the AIDS virus - so it was a non-issue.

Canberra, Australia was a different case altogether. All of the Marines had girlfriends there - and some of them had more than one. It was a situation ripe for disaster. Even though I made a concerted effort to make as many unannounced visits to the Marines' quarters as possible, and made it quite clear that I had a zero tolerance policy for transgressors, I knew more had to be done. I simply couldn't be there all of the time. The solution I decided upon was to make my A-Slash (Assistant Detachment Commander) personally responsible for any and all violations, and to his credit there were none (that I know of).

Despite my stand on the enforcement of battalion regulations, no one can ever say I didn't encourage my Marines to have a good time within the boundaries of the rules. Parties were thrown at the Marine House on a monthly basis, and some of them were quite memorable. They were so good, in fact, that we began to receive regular invitations to attend functions hosted by the Australian military and police in return, and we ended up developing a very cordial relationship with each of those organizations.

One incident that stands out took place early in my tour in Canberra. It was the tail end of one of our monthly parties, and as the crowd thinned out I noticed one of those who remained was an Australian girl who worked in the embassy. I had met her on a couple of occasions and she had impressed me as being quite intelligent and witty, but unfortunately that was not the case on this particular evening. She was, as they say, a bit "green around the gills." When it came time for her to go home it became apparent she could not even walk, so it became necessary for me to carry her outside to the cab her friend had called. When I placed her in the back seat I thought that was that, but I was wrong.

I felt bad for that poor girl, I really did - but if I had known what she had done I probably wouldn't have been quite so sympathetic. During the course of the party she had been drinking shots poured for her by one of my Marines, who was no doubt doing his level best to get her drunk. He succeeded. When all of that liquor kicked in that nice girl decided she needed to locate a bathroom in order to perform what the Aussies sometimes refer to as the "technicolor yawn." She somehow managed to get there alright, but couldn't find the light switch – so instead she just felt around in the dark for something porcelain with a lid, and that's where she proceeded to deposit the contents of her stomach.

Well, the following morning one of my Marines went downstairs to do a load of laundry, and you can imagine his surprise when he opened the lid of the washing machine to discover the mess left behind by our guest from the night before. When that sweet girl was informed she had mistaken the washer for the toilet she was mortified, and promptly sent a bouquet of flowers to the Marine who had made the discovery (and who had cleaned up the mess) along with a written apology. But I guess in the end the impression she made wasn't all that bad, since she ended up marrying my A-Slash a couple of years later!

MELVIN!

"If you obey all the rules, you miss all the fun."
– Katharine Hepburn

What do you do if you are a thirty-two-year-old Marine Captain, and you've been out with the boys, and the next morning you can't remember where you left your car? Well, Mac, if your name is Johan S. Gestson, here is what you do. First, you take a Bromo. Then you go back to the scene of the crime - the Sandpiper Bar in Laguna Beach, California - order up a serving of clear broth (four parts gin, one part vermouth), and ask the others if they remember. They can't of course. Better call someone!

"I'll call someone in Peoria, maybe they'll know," you say, with the fine luminescent logic of a Marine. You've never been to Peoria, of course, and you don't know anyone there, and it's two thousand miles away, but what the hell. "Get me Harry Miller in Peoria, Illinois," you tell the operator. The line is busy? "Well, get me Melvin Miller, then." And lo, there *is* a Melvin Miller in Peoria, and the operator gets the fellow on the phone.

Poor Melvin. He will never forget that moment during last Labor Day weekend when the first phone call from Laguna Beach came in. A hefty, amiable foreman of the Caterpillar Tractor Company, Miller was painting his house on Peoria's west side at the time. Puzzled and vaguely apprehensive, Mel Miller wiped the paint from his hands and took the person-to-person call from California.

"I've lost my car," said Captain Joe Gestson without preamble. "Have you seen it around?"

"Who is this, and where are you?" Miller asked.

“I’m Joe Gestson, and I’m in the Sandpiper Bar in Laguna Beach,” the Captain replied. “It’s a canary yellow 1950 Chevrolet convertible... lost it just last night.”

“I’ll sure keep an eye out for it,” Mel Miller said pleasantly. “I’ll let you know if it turns up.”

Joe went back to the bar and told his pals about Melvin Miller. Then someone came in and told him the car was in front of a Laguna Beach restaurant.

It wasn’t until two AM the next morning that Joe remembered Melvin Miller in Peoria. Why, the poor fellow was probably still out looking for the car. He put in another call and roused Melvin from a sound sleep. “Just wanted to tell you that I found the car and not to keep looking for it,” said Joe.

“Call anytime I can be of service,” said Mel, climbing back into bed.

It was more than a month later, and past midnight, when the ice ran out during a little officer’s party. “We’ll call down to the store,” someone said. “Wait a minute,” said Joe, his dialing finger at the ready. “I’ll call Melvin Miller first.”

“Got any ice?”Joe asked when Melvin answered drowsily. Mel said he did. “Could we borrow some?” Joe asked. “Sure,” Mel said. “Who is this?”

“Joe Gestson, in Laguna Beach,” Joe said. “I can’t send ice there,” Mel protested. “It’s simple,” said Joe. “Pack it in sawdust and shoot it right along.”

“OK, pal,” said Mel. “Get some sleep.”

Soon the idea spread around Camp Pendleton that the best thing a Marine could possibly do in the middle of the night was call Melvin Miller in Peoria - about anything, or nothing at all. Marines talked to him about errant girlfriends, corn growth in Peoria, the rainfall in Illinois. “I never get mad,” said Mel, who was in the Navy during World War II. "I just light up my pipe, sit down, and talk to them."

A friend of Joe's, Lieutenant Bob Neal, was driving from Camp Pendleton to Eglin, Illinois one day, so he stopped by to see Melvin, had a chat, took a picture, and sent it back to California. Up went the picture behind the Sandpiper's bar, on an election poster proclaiming: "We're for Melvin Miller!"

Somebody decided Mel had to be brought to Laguna, so a howitzer shell was set up on the bar as a receptacle for the "Melvin Miller Fund."

"We're for Melvin Miller!" hollered Marines to startled tourists. "Melvin Miller" signs popped up all over Laguna Beach and on barracks walls. Blue pennants, with "Melvin Miller" in gold letters, and T-shirts announcing "Melvin Miller Week, Sept 1-5," found a booming market. By the end of the week, with more than three hundred dollars in the Melvin Miller fund and more pledged (keep in mind it was 1960), Miller's Labor Day anniversary celebration was assured.

Joe Gestson had cleared things with the Caterpillar Company to get Miller the time off. An airline had promised the red-carpet treatment for him. A chartered bus, jammed with his admirers, met him at the airport and took him to the Coast Inn, where a special beach chair and umbrella with his name on them awaited below the windows of his suite on the Pacific shore.

September 1st was designated "Melvin Miller Day" at the Del Mar Racetrack, with a race and a purse in his honor. A buffet testimonial dinner was held at the Sandpiper that night. There was a luncheon with movie stars at Twentieth Century Fox Studios, a trip to the bullfights in Tijuana (with a matador hat for Mel and a bull dedicated to him), swimming parties, cocktail parties, luncheons and dinners.

And then? Then Melvin Miller went back to Peoria to wait for the phone to ring again. Just dial Peoria 4-9898 – if you're a Marine!

This story originally appeared in Newsweek Magazine in August of 1962

AN UNLIKELY HERO

By Rebecca Liss

"I say to our enemies, God may show you mercy. We will not." – Senator John McCain, 12 September 2001

Those of you who have read 'Swift, Silent and Surrounded' know the story of Sergeant Major Mike Curtin, how he recovered the remains of a Marine killed in the Oklahoma City bombing, and was later recovered himself by other Marines from the rubble of the World Trade Center in the aftermath of 9/11. What you probably don't know is there was yet another Marine who responded to the greatest single disaster in U.S. history, and whose presence also made a real difference:

Only twelve survivors were pulled from the rubble of the World Trade Center after the towers fell on September 11, despite intense rescue efforts. Two of the last three to be located and saved were Port Authority police officers. They were not discovered by a heroic firefighter, or a rescue worker, or a cop. They were discovered by Dave Karnes.

Karnes hadn't been near the World Trade Center. He wasn't even in New York when the planes hit the towers. He was in Wilton, Connecticut, working in his job as a senior accountant with Deloitte Touche. When the second plane hit, Karnes told his colleagues, "We're at war." He had spent twenty-three years in the Marine Corps infantry and felt it was his duty to help. Karnes told his boss he might not see him for a while. Then he went to get a haircut.

The small barbershop near his home in Stamford, Conneticut was deserted. "Give me a good Marine Corps squared-away haircut," he told the barber. When it was done, he drove home to put on his uniform. Karnes always kept two sets of Marine fatigues hanging in his closet, pressed and starched. "It's kind of weird to do, but it comes in handy," he said. Next Karnes stopped by the storage facility where he kept his equipment – he'd need rappelling gear, ropes, canteens of water, his Marine Corps K-Bar knife, and a flashlight, at least. Then he drove to church. He asked the pastor and parishioners to say a prayer that God would lead him to survivors. A devout Christian, Karnes often turned to God when faced with decisions.

Finally, Karnes lowered the convertible top on his Porsche. This would make it easier for the authorities to look in and see a Marine, he reasoned. If they could see who he was, he'd be able to zip past checkpoints and more easily gain access to the site. For Karnes, it was a "God thing" that he was in the Porsche - ironically a Porsche 911 - that day. He'd only purchased it a month earlier - it had been a stretch, financially. But he decided to buy it after his pastor suggested that he "pray on it." He had no choice but to take it that day because his Mercury was in the shop. Driving the Porsche at speeds of up to 120 miles per hour, he reached Manhattan - after stopping at McDonald's for a hamburger - in the late afternoon.

His plan worked. With the top off, the cops could see his pressed fatigues, his neatly cropped hair, and his gear up front. They waved him past the barricades. He arrived at the site – "the pile" - at about 5:30 PM. Building Seven of the World Trade Center, a forty-seven story office structure adjacent to the fallen twin towers, had just dramatically collapsed. Rescue workers had been ordered off the pile - it was too unsafe to let them continue. Flames were bursting from a number of buildings, and the whole site was considered unstable. Standing on the edge of the burning pile, Karnes spotted another Marine dressed in camouflage. His name was Sergeant Thomas. Karnes never learned his first name, and he's never come forward in the time since.

Together Karnes and Thomas walked around the pile looking for a point of entry farther from the burning buildings. They also wanted to move away from officials trying to keep rescue workers off the pile. Thick, black smoke blanketed the site. The two Marines couldn't see where to enter. But then "the smoke just opened up." The sun was setting and through the opening Karnes, for the first time, saw clearly the massive destruction. "I just said 'Oh, my God, it's totally gone.'" With the sudden parting of the smoke, Karnes and Thomas entered the pile. "We just disappeared into the smoke - and we ran."

They climbed over the tangled steel and began looking into voids. They saw no one else searching the pile - the rescue workers having obeyed the order to leave the area. "United States Marines!" Karnes began shouting. "If you can hear us, yell or tap!"

Over and over, Karnes shouted the words. Then he would pause and listen. Debris was shifting and parts of the building were collapsing further. Fires burned all around. "I just had a sense, an overwhelming sense come over me that we were walking on hallowed ground, that tens of thousands of people could be trapped and dead beneath us," he said. After about an hour of searching and yelling, Karnes stopped.

"Be quiet," he told Thomas, "I think I hear something."

He yelled again. "We can hear you. Yell louder." He heard a faint muffled sound in the distance.

"Keep yelling. We can hear you." Karnes and Thomas zeroed in on the sound.

"We're over here," they heard.

Two Port Authority police officers, Will Jimeno and Sergeant John McLoughlin, were buried in the center of the World Trade Center ruins, twenty feet below the surface. They could be heard but not seen. By jumping into a larger opening, Karnes could hear Jimeno better. But he still couldn't see him. Karnes sent Thomas to look for help. Then he used his cell phone to call his wife, Rosemary, in Stamford and his sister Joy in Pittsburgh. (He thought they could work the phones and get through to New York police headquarters.)

"Don't leave us," Officer Jimeno pleaded. He later said he feared Karnes' voice would trail away, as had that of another potential rescuer hours earlier. It was now about 7 PM,. and Jimeno and McLoughlin had been trapped for roughly nine hours. Karnes stayed with them, talking to them until help arrived in the form of Chuck Sereika, a former paramedic with an expired license who had pulled his old uniform out of his closet and come to the site. Ten minutes later Scott Strauss and Paddy McGee, officers with the elite Emergency Service Unit of the NYPD, also arrived.

The story of how Strauss and Sereika spent three hours digging Jimeno out of the debris, which constantly threatened to collapse, has been well told in the New York Times and elsewhere. At one point, all they had with which to dig out Jimeno were a pair of handcuffs. Karnes stood by, helping pass tools to Strauss, and offering his Marine K-Bar knife when it looked as if they might have to amputate Jimeno's leg to free him. (After Jimeno was finally pulled out, another team of cops worked for six more hours to free McLoughlin, who was buried deeper in the pile.)

Karnes left the site that night when Jimeno was rescued and went with him to the hospital. While doctors treated the injured cop, Karnes grabbed a few hours sleep on an empty bed in the hospital psychiatric ward. While he slept, the hospital cleaned and pressed his uniform.

On the anniversary of the attack and the rescue, officers Jimeno and Strauss were part of the formal "Top Cop" ceremony at the New York City Center Theater. Earlier the two appeared on a nationally televised episode of *America's Most Wanted.* Jimeno and McLoughlin also appeared on the *Today* show. They are heroes.

Karnes is a hero, too. But it's also clear he is a hero in a smaller, less national, less public, less publicized way than the cops and firefighters are heroes. He's hardly been overlooked - the program I work for, 60 Minutes II, interviewed him as part of a piece on Jimeno's rescue - but the great televised glory machine did not pick him. Why? One reason seems obvious - the cops and firefighters are part of big, respected, institutional support networks. Americans are grateful for the sacrifices their entire organizations made. Plus, the police and firefighting institutions are tribal brotherhoods. The firefighters help and support and console each other; the cops do the same. They find it harder to make room for outsiders like Karnes (or Chuck Sereika). And, it must be said, at some macho level it's vaguely embarrassing that the professional rescuers weren't the ones who found the two survivors. While the pros were pulled back out of legitimate caution, the job fell to an outsider who drove down from Connecticut and just walked onto the burning pile.

Columnist Stewart Alsop once famously identified two rare types of soldiers, the "crazy brave" and the "phony tough." The professionals at Ground Zero - I interviewed dozens in my work as a producer for CBS - were in no way phony toughs. But Karnes does seem a bit "crazy brave." You'd have to be slightly abnormal - abnormally selfless, abnormally patriotic - to do what he did. And some of the same qualities that led Karnes to make himself a hero when it counted may make him less perfect as the image of a hero today.

Officer Strauss tells a story that gets at this. When he was out on the pile, trying to pull Officer Jimeno free, Strauss shouted orders to his volunteer helpers – "Medic, I need air," or "Marine, get me some water." At one point, in the middle of this exhausting work, Strauss asked if he could call them by their names to facilitate the process. The medic said he was "Chuck."

Karnes said, "You can call me 'Staff Sergeant.'"

"That's three syllables!" said Strauss, who needed every bit of energy and every second of time. "Isn't there something shorter?"

Karnes replied. "You can call me 'Staff Sergeant."

The story of Jimeno and Strauss' rescue was told onscreen in the 2006 Oliver Stone film *World Trade Center* starring Nicholas Cage. After hearing about the movie the mysterious "Sergeant Thomas" stepped forward and identified himself as Jason Thomas.

LEGEND OF THE CUBI CAT

By Doug Talley

"Enjoy yourself – these are the 'good old days' you're going to miss in the years ahead."

Naval aviators are a strange bunch, even more so than regular pilots. It takes a special type of person to land a jet aircraft on a bobbing, floating runway in the middle of the night, and it is not surprising to discover the personalities of those pilots are a bit, shall we say, eccentric.

The dangers associated with naval aviation can be pretty stressful, and pilots have found some pretty ingenious and entertaining ways to blow off some of that stress. A good example is the 'Carrier Qual,' which involves a simulated landing on a beer soaked table into the headwind created by a fan, with the object being to snag the "arresting cable" with your feet before running out of "flight deck." But of course there is always room for improvement, as was demonstrated by some inventive jet jocks during the Vietnam War:

If you're old enough to have served in the Navy or Marine Corps during the Vietnam War, and particularly if you were an aviator, chances are you've heard of the infamous "Cubi Point Catapult." Cubi Point Naval Air Station and the adjoining Subic Bay Naval Base in the Philippines was a place where war-weary Marines and sailors could let off a little steam after flying combat missions over Vietnam or spending weeks on the gunline aboard ships on Yankee Station.

The managers of the Cubi Point Officers' Club, as well as their counterparts at the other officer and enlisted clubs, were forever tasked with devising new and challenging ways of keeping the warriors entertained. Enter Commander John L. Sullivan and the now famous Cubi Point Officers' Club catapult.

The catapult came into existence in 1969 and immediately created a division within naval air among those who had ridden the cat and caught the wire, and those who had ridden the cat, missed the wire and gotten soaked. The escapades of Navy and Marine pilots at the Officers' Club is the stuff of legend.

"These tales will be handed down and embellished as long as we have aircraft carriers in that part of the world," Sullivan said in an article he wrote for *Wings of Gold* magazine. One of these escapades involved catapulting a squadron mate down a half dozen stairs in a chair from the bar upstairs onto the dance floor below.

"The fact the chair had castors helped little on the stairs. Rarely did a pilot make it down the stairs and onto the dance floor in an upright posture. Most arrived in a crumpled mess."

"There were broken bones, severe strains, small concussions and numerous other injuries that grounded crack combat pilots," former Commander in Chief, Pacific Fleet, Admiral Maurice 'Mickey' Weisner said in a phone interview. Weisner said that he and Vice Admiral Ralph Cousins, commander, Task Force-77, suggested to Captain 'Red Horse' Meyers, NAS Cubi Point, that the chair catapulting be eliminated because of the injuries. At the time, Sullivan was the Aircraft Intermediate Maintenance Department (AIMD) officer.

"I was called to the skipper's office and asked to come up with a solution," Sullivan recalled. "After a great deal of consultation with my maintenance officers we realized we had an excellent window of opportunity. A new lower club extension to replace an old bamboo bar was in progress. From that point on we let our imaginations run wild."

Heading off to the surplus yard, Sullivan and his band of AIMD scavengers liberated a banged up refueling tank which was quickly converted by the metal smiths into something resembling an A-7 Corsair II. The 'aircraft,' Sullivan recalls, was six feet long had shoulder straps and a safety belt and was equipped with a stick that, when pulled back sharply, released a hook in the rear of the vehicle to allow arrestment. Propulsion was provided by pressurized nitrogen tanks hooked up to a manifold.

"This arrangement provided enough power to propel the vehicle to fifteen mph in the first two feet," said Sullivan. "Acceleration of zero to fifteen mph in two feet is the equivalent of the G force of World War II hydraulic catapults. We named the vehicle 'Red Horse One' in honor of our skipper, Captain Meyers."

Beyond the exit from the club was a pool of water three and a half feet deep. Each pilot had six inches to play with if he was to make a successful arrestment. Successful pilots, according to the commander, were held in high esteem by their peers, and their names were inscribed in gold letters on the club's Wall of Fame.

Reaction time was short, because the wire was some fourteen feet from the nose of the vehicle. The downward curvature of the track had to be precise, because the rollers would bind if the curvature were too sharp. Since the pool water was the force that stopped the vehicle, it was also important to get the vehicle as deeply and as quickly into the pool as possible - so engineers from the Strategic Aircraft Repair Team used their 'slip sticks' to solve the problem. The vehicle was retrieved from the water by a mechanical winch and cable connected to an eye welded to the back of the "A-7."

Sullivan said that Rear Admiral Roy Isaman, (Naval Air Test Center commander, 1971-74), had a bronze plaque made in Hong Kong which was bolted to the wall next to the catapult with the inscription, 'Red Horse Cat-House.'

"The first night the catapult was in operation it attracted a huge crowd, and Rear Admiral Isaman was the first to ride the vehicle after it was declared safe by the BIS (Board of Inspection and Survey). No problem, since I had recently arrived from the test center at Patuxent River and was declared the BIS representative," Sullivan recalled. "Rear Admiral Isaman manned the cockpit, saluted and was launched. He dropped the hook early and we awaited the hook skip but it didn't happen. Instead the hook caught the rubber we had attached to the steel bumper short of the wire. The hook tore the rubber from the bumper and caught the wire. To the howl of the disappointed junior officers, there was no wet admiral this time. Isaman became the first pilot to trap in the vehicle." After being presented with a bottle of champagne, Isaman's name was then enshrined on the 'Wall of Fame.' Some forty pilots rode the Cat that night before another successfully trapped," Sullivan laughed.

Word of the Cat quickly spread throughout Southeast Asia, and even attracted Air Force F-4 pilots from Clarke AFB. "They would come swaggering in loudly claiming they were equal to the task. Each and every one of them failed to catch the wire, much to the delight of the Navy onlookers."

"Enlisted men from AIMD operated and maintained the catapult during their off time. They were compensated for their work from funds we took in for the operation of the Cat. It cost nothing to ride the Cat," Sullivan emphasized, "providing they caught the wire. However, it cost five dollars if the rider required rescue from the pool." Sullivan said that of the many dignitaries who attempted to ride the cat, his favorite was Under Secretary of the Navy John Warner (later a U.S. Senator from Virginia). "After flying in from Japan the Secretary was taken to the club for lunch by Rear Admiral Isaman and Captain Meyers. The secretary had heard of the Cubi CAT and unhesitatingly requested to ride it. Captain Meyers looked at me; I nodded and immediately took steps to get a crew ready. Word spread rapidly that Under Secretary of the Navy John Warner would try his luck. The club was soon packed with onlookers. Before launch we outfitted the secretary in a set of white linen coveralls with 'Red Horse Cat House' embossed in bright red letters on the back. Amid the cheers of the onlookers, the secretary bravely launched and promptly landed in the pool. We catapulted him five times after that and each time he got wet. The skipper kicked the bumper plate back about an inch each time hoping he would catch the wire. While the official never noticed this, we all did. He told the skipper after his fifth trip into the pool, 'it can't be done.' By this time the bumper was back some twelve inches from the wire and was an easy arrest for a pilot who had a launch or two on the CAT under his belt. So 'Red Horse,' in his tropical whites, strapped in. Before launch one of the junior officers kicked the bumper forward to its original six-inch position. Meyers launched and to the delight of the visiting official, settled ignominiously into the pool. Secretary Warner wouldn't take off the coveralls. He and the skipper, both wringing wet, sat down to lunch with dry colleagues. Several hours later, still wearing the coveralls, the secretary boarded his aircraft. The tale of his Cat adventures would be told at the Pentagon, he informed us, and the coveralls would be testimony to the validity of his tale."

Sullivan completed his tour at Cubi Point in 1971 and returned to Patuxent River. "I am happy to say there were no injuries from riding the Cat during that period… only wounded pride," he said.

Sullivan returned to Cubi Point in 1979 while in the employ of Grumman Aerospace Corporation as the Project Manager for the C-2 COD. Much to his dismay the Cat was gone. "The tracks were covered, and the pool was filled with cement. When I was introduced to the new club manager, he asked if I could assist him in putting in a new Cat. I felt like a dinosaur whose time had passed. I always believed that as long as there was a Cubi Point there would be a fun place for naval aviators to unwind. In the midst of it all would be the 'Cat' and the 'Wall of Fame.' Now both are gone. What remains is my fond memories of the officers and men of AIMD whose ingenuity and hard work made the 'Cat' a reality in 1969. Today it remains a 7th Fleet legend."

A GRUNT'S TALE

By Richard W. Williams

"Know your enemy and know yourself." – Sun Tzu

I was a grunt in India Company 3/5 in 1969. But this is not war story. This is a story about the Espirit de Corps of the 1st Marine Division. I lived in Boca Raton, Florida prior to joining the Marines in 1968, and learned that there was a Marine who lived close by my home. One day I knocked on his door and his wife answered. I merely said I was considering joining the Marines, and that I understood her husband was a former Marine. I was hoping he'd let me ask a few questions about what to expect.

Like any good Marine's wife she let me in and introduced me to her husband, "Archie." Archie was quite old. He sat in his winged-back chair with a quiet repose, and in spite of his failing eyesight he fixed me with a steady gaze, politely smiled and simply said, "Welcome aboard."

We talked the afternoon away. Archie patiently answered my questions about the Marine Corps, Parris Island and careers in the Corps. All he related to me about his exploits in the Marines was that he loved the Corps and every minute he had served in it. As the late afternoon sun dipped on the horizon I bid him farewell, and promised I would return to see him after I finished Boot Camp.

I kept my promise and visited him nearly every day I was back from Parris Island. He and his wife were gracious hosts. As I sat and learned from Archie, his wife would serve us brandy in the afternoon to go with the cigar Archie enjoyed only once a day. I felt extremely bonded with Archie for sharing his ritual with me.

As my leave drew to a close and I prepared to go to Vietnam, Archie's and my conversations grew deadly serious. He gave me tip after tip on how to fight and even how to win campaigns. As an enlisted snuffie, I didn't think the High Command would be interested in my opinions on running a campaign, but I listened in utter fascination to Archie's knowledge. He told me what to expect in war, and what not to fear.

After his brandy one evening he said, "Don't worry if you are ready for the task of war. Because no sane man is ever ready. There is only one thing that makes a good warrior and that is a man who cares for his fellow man. That is why the Marines do so well at making war. We respect each other. We'd rather die than to let down our comrades. You see, there are many reasons a young man marches off to war - patriotism, duty, honor, adventure; but only one reason he actually fights once he is in a war. He fights for the men next to him. Marines don't endure the hell of combat for any lofty principles. Marines fight because each Marine acknowledges the loftiest principle of all: he acknowledges and accepts the responsibility of being his brother's keeper. That's why you will fight. You are a Marine and you will protect your unit at all cost."

Archie asked me to write and keep him abreast to what I experienced in Vietnam. He gave me his address. I thanked him and promised I would write as soon as I landed and found out what my FPO address would be. Without looking at it, I folded the paper and put his address in my wallet and marched off to war.

Naturally, I lost his address. However, I sent a letter to him through my father letting him know I had landed and providing him with my FPO. I had been in Vietnam less than a month when I got a response from Archie. He simply asked me to tell him how we were conducting the war, what were my impressions. The name on the return address was General Archer A. Vandegrift, USMC (Ret). My friend Archie was the former Commanding General of the 1st Marine Division! He was awarded the Medal of Honor at Guadalcanal and later became Commandant of the Marine Corps. At PI, we had learned all about General Vandegrift. But being as dumb as a box of rocks, I never really remembered "Archie's" last name until his letter arrived. I just remembered it was "Van" something. I sat down in the sweaty jungle rot and stench and began what would be a long series of letters from one snuffie to the ex-Commandant and most famous CG of my Marine Division.

I started it out simply. "Dear General Vandegrift, Vietnam is like a large island where the enemy has kept a seaway open. The enemy also has a secret weapon. The seaway is the Ho Chi Minh trail. The secret weapon is their ability to resupply themselves using technologies that have existed since the stone age. We ignore the seaway, leaving it open and try to use high technology to cause collateral damage to their stone-age production capacity. It's like dropping firecrackers on ants. So the enemy will continually be resupplied. And we will continually be resupplied. That means this fight will go on until one side or the other tires of it. On the ground, your Marines are just that, Marines. We are doing just what you predicted, fighting for the guy next to us.

Other than that it don't mean nothing, but what does mean something is that for all those months you never let on who you were. It was just two Marines, no rank. That's why I serve, because of men like you who have made the Marine Corps something worthy to fight for. Semper Fidelis."

The General wrote back and agreed that an enemy must be denied resupply. A war of attrition is less costly to a Third World country then it is to a high-tech country. He said that the bombing and blockading of Haiphong Harbor and an end run up the Ho Chi Mihn trail coupled with a staggered attack due north would end this war in a few months. But, without a Pearl Harbor, the American people don't have a heart for war. That was America's greatest strength, he said. We only like to fight when we are mad. And, when we are mad we fight like no other civilization in the history of the world.

This story isn't about famous people I have known. I was then, and am now, nobody - just a simple grunt. But the most famous CG of the 1st Marine Division sitting down and talking to a lowly private just shows what the Marine Corps is made of. It shows that the Corps' motto, Semper Fidelis, is more than mere words. It is a way of life.

THE GREEN MONSTER

By Charles R. "Chuck" Dowling

"All warfare is based upon deception." - Sun Tzu

Back in 1958, when I checked into my first duty station in Twenty-nine Palms, California as a newly minted Second Lieutenant, I was assigned to one of the 105-mm artillery batteries in the First Field Artillery Group. The 1st FAG, a name many were leery of owning up to, was part of Force Troops. Force Troops was composed of our heavy artillery group, the 1st MAAM battalion, (the first ground to air missile systems in the Corps), some twin-forty units and a few other air defense units.

The artillery batteries in the Field Artillery Group were constructed a little differently that the division artillery Regiments, like the Tenth, Eleventh and Twelfth Marines. Each FAG battery was structured to be able to operate independently of support elements. As a consequence, in addition to our artillery pieces, we also had our own motor transport section with our own trucks and repair capabilities, our own supply section and our own communication section. In personnel and equipment, our batteries were nearly double the size of a traditional artillery battery.

As a result of this size and the number of artillery pieces and vehicles involved, the FAG gun and motor parks were spread out over a large area. The number of guards needed to provide security for the unit was substantial, and posting and inspecting guards around the perimeters of the FAG complex was a big operation. One Marine standing guard would have to walk four or five miles in a four hour shift, just to make sure he was able to cover his post properly. Changing and supervising the guard required a vehicle.

Shortly after my arrival I, along with a couple of other newly arrived Lieutenants, were placed on the 1st FAG duty roster and within two weeks I was serving my first tour as Officer of the Day.

When we posted the guard, I rode with the Sergeant of the Guard to learn the lay of the land. We would then alternate inspection tours during the night, ensuring that each of us could get at least a few hours of unbroken sleep. There was also a Corporal of the Guard and a duty driver. For the Corporal, his job was to remain awake all night, to take calls, check in new arrivals etc. The duty driver also remained awake all night, since his job was to transport the Sergeant of the Guard or the OD to wherever they had to go. They would usually be allowed to sleep the next day. The Sergeant of the Guard and the Officer of the Day had to return to their regular duties however. No rest for the weary.

When the Sergeant and I returned from posting the guard we set up our inspection rotation. I don't remember the exact times, but one of the tours I was to take was about 0200 or 0300. The Corporal of the Guard woke me up at the correct time, and off I went with the duty driver.

Then, around the halfway mark though our rounds, we heard a rifle shot. It came from the most eastern part of our area. That part of the parking area was at the extreme edge of 1st FAG's gun parks. From this point on the terrain was wild desert, gently sloping up to some craggy and rocky sun scorched hills.

Needless to say, I was startled. This was peacetime. No wars were going on (except for the Cold War). Who was doing the shooting? I was imagining the worst. I told the driver to load his weapon. I loaded mine. Then we raced to the spot where we thought we had heard the shot.

When we arrived all was quiet. We looked for the guard, and finally found him lying under one of the trucks, peering out into the desert. He yelled at us to cut the lights of the jeep. We complied. Then I crab-walked over to him. He was in a prone position, with his rifle tucked into his shoulder. He was watching the desert area beyond the park boundary intently. His face glistened with sweat.

"Sir, there's something out there. It came at me. It was green… I shot it. It went away." Many words bubbled out, but that was the gist of it. I didn't know what to make of the story or of the Marine. I looked to where he was looking, but saw nothing. I vaguely remember now that although it was nighttime, the landscape could be seen quite well. We waited for awhile and then the three of us gingerly approached the area where the Marine had said he saw the green thing. We found nothing.

In the Marine Corps, firing a weapon other than on the weapons range or at the appropriate time has to be explained. I'm sure the other services have the same or similar requirements. Thus an investigation was launched.

The investigation went on for some time, but turned up nothing of interest. The Marine was removed from duty for awhile and underwent some interrogation, but then was released back to active status. The incident was forgotten. Then it happened again, about four or five months later.

From this point on the story of the mysterious incident comes to me third hand. I had no personal involvement. This time it was another Marine who fired the weapon, but it was in the same area where the first incident took place and the Marine was in the same platoon as the first. This Marine identified the intruder as a "Green Monster." He said it glowed and moved slowly and smoothly toward him. He said he called out for it stop, and then fired. Immediately the green glow disappeared. He then ran back to headquarters to report the happening.

Obviously another investigation ensued, again finding nothing. This Marine was sent to "Office Hours" for leaving his post. He received some extra duty as punishment. He also received a thorough interrogation and was eventually returned to normal status.

The third incident took place about a month later. This time it involved the first Marine again, a man I will call 'Private Mahon.'

Mahon - I don't have a first name - was, in the vernacular of the day, "a piece of work." He had kind of a fire plug build, square head, big neck and a short blocky body. His hair, what their was of it, was a dirty blonde. He had already built a legend of sorts for himself at the time of the "Green Monster" incidents.

A few months before, for some reason known only to the Gods of the Marine Corps, an order had come down from on high to inventory all the light poles on the base. Each pole had a number. The number was embossed on a metal plate about the size of a dog tag and nailed to the pole. And so six-man work party was formed including Mahon, and they were dispatched around the base to "get the numbers off the light poles."

Now that order could be interpreted in a number of different ways. Five of the members of the work party came back with a list of the numbers in their assigned areas. Mahon came back, long after the others, with a pile of numbered plates that he had pried off the light poles in his assigned area. Needless to say, there was a great deal of irritation over this among the supply people.

In another incident, during a live fire exercise, Mahon's 105-mm howitzer section received an order to break off from their indirect fire mission and turn in a different direction and then fire their howitzer as a direct fire weapon. The exercise was to simulate a combat situation, whereby an enemy may get close to an artillery battery and attack its position with infantry. The end result would be like some of the Civil War battles where cannons fired directly into charging troops with grapeshot.

For the "Cannon Cockers," the mission involved a great deal of physical effort. They would literally have to manhandle the howitzer to a new firing position. Well all hands turned to, except Mahon. He just watched. At the point where the struggling Marines were lifting the weapon's trails to swing it around the section chief, a grizzled Staff Sergeant, muscles bulging as he helped his men, looked up and saw Mahon simply standing there. He roared at him. "Mahon, you m----- ----ing ----bird, get on that ----ing trail!." (Expletives deleted).

Mahon immediately ran over to the gun and jumped up onto the trail, his weight knocking it out of the hands of the struggling gun crew, forcing some to their knees. Needless to say the Section Chief, aided by a few members of the gun crew, immediately administered some "field" discipline to Private Mahon.

The incident made it to "Office Hours" too, but went no further. What Mahon did was not a legally punishable incident. What the Section Chief and the others did could have been. In any event, Mahon's reputation was established.

The way the third "Green Monster" incident seems to have unfolded was little different from the other two. There was no shooting. In the middle of a clear starry night the guard walking the post just disappeared. Mahon was the guard. His post, this time, had been near the place of the first two incidents, but not at exactly the same spot.

When it was discovered that one of the guards was missing, the remainder of the guard was turned out and a search ensued. Then the base provost martial and the MPs pitched in. The next day more troops joined the search. Mahon was not found, but the searchers did discover a cartridge belt in an area about a half-mile into the desert where a number of desert shrubs had been dug up or broken. There were no cartridges in the cartridge belt. The ground itself showed signs of what appeared to be a struggle. There were no footprints or tire tracks, but the gravelly sand had been trampled and actually dug up in a few places.

There was no way to tell if the cartridge belt was Mahon's or not, but he would have been wearing a similar belt while on guard duty. I left the base to go to a school around that time, and so I lost track of the incident. I knew an even more extensive search had since been made, with no results. When I returned, for the most part it seemed to have been forgotten. Of course once in awhile someone would talk about "The Green Monster," but no one could add anything new. The story just drifted off into legend.

WELCOME HOME

Author Unknown

"You can curse the darkness, or you can light a candle."

It wasn't long ago that troops returning home from Vietnam were spat on in airports and called baby killers. Remember that? In the 1960's and early 1970's it was en vogue to express contempt for the military, and I don't think it was a coincidence many of those expressing that contempt were also experimenting with LSD and other mind altering drugs. Those people were completely out of touch with the real world.

Times have changed. These days even the most outspoken of liberals will not openly criticize the military – instead they camouflage their contempt by directing their ire at political targets while publicly "praising" those in uniform. Much of the credit for this turnaround must go to President Ronald Reagan. It was he who made us proud once again – proud to be Americans, and proud to serve our country. his story demonstrates the legacy he has left us:

I sat in my seat of the Boeing 767 waiting for everyone to hurry and stow their carry-ons and grab a seat so we could start what was sure to be a long, uneventful flight home. With the huge capacity of the plane and slow moving people taking their time to stuff luggage far too big for the overhead and never paying much attention to holding up the growing line behind them, I simply shook my head knowing that this flight was not starting out very well. I was anxious to get home to see my loved ones so I was focused on "my" issues and just felt like standing up and yelling for some of these clowns to get their act together. I knew I couldn't say a word, so I just thumbed thru the "Sky Mall" magazine from the seat pocket in front of me. You know it's really getting rough when you resort to looking at that overpriced, useless sky mall crap to break the monotony. With everyone finally seated, we just sat there with the cabin door open and no one in any hurry to get us going although we were well past the scheduled take off time. "No wonder the airline industry is in trouble," I told myself.

Just then, the attendant came on the intercom to inform us all that we were being delayed. The entire plane let out a collective groan. She resumed speaking to say, "We are holding the aircraft for some very special people who are on their way to the plane and the delay shouldn't be more than five minutes. The word came after waiting six times as long as we had been promised that "I" was finally going to be on my way home. Why the hoopla over "these" folks? I was expecting some celebrity or sport figure to be the reason for the hold up. "Just get their butts in a seat and let's hit the gas," I thought.

The attendant came back on the speaker to announce in a loud and excited voice that we were being joined by several U. S. Marines returning home from Iraq! Just as they walked on board, the entire plane erupted into applause. The men were a bit taken by surprise by the 340 people cheering for them as they searched for their seats. They were having their hands shook and touched by almost everyone who was within an arm's distance of them as they passed down the aisle. One elderly woman kissed the hand of one of the Marines as he passed by her. The applause, whistles and cheering didn't stop for a long time.

When we were finally airborne, I was not the only civilian checking his conscience as to the delays in "me" getting home, finding my easy chair, getting a cold beverage and having the remote in my hand. These men had done for all of us and I had been complaining silently about "me" and "my" issues. I took for granted the everyday freedoms I enjoy and the conveniences of the American way of life, and I took for granted that others had paid the price for my ability to moan and complain about a few minutes delay so those Heroes could go home to their loved ones.

I attempted to get my selfish outlook back in order and minutes before we landed I suggested to the attendant that she announce over the speaker a request for everyone to remain in their seats until our heroes were allowed to gather their things and be first off the plane. The cheers and applause continued until the last Marine stepped off and we all rose to go about our too often taken for granted everyday freedoms. I felt proud of them. I felt it an honor and a privilege to be among the first to welcome them home and say "Thank You" for a job well done. I vowed that I will never forget that flight nor the lesson learned. I can't say it enough – thank you!

CALL SIGNS OF THE TIMES

"Forgive your enemies, but never forget their names."
- John F. Kennedy

One of the really cool things about talking on the radio is you get to have a call sign. You know, like "Maverick," "Gunfighter" or the ever popular "Warlord." If you get to pick your own it will almost certainly be something very flattering and warlike - but if it gets picked for you it will probably be just the opposite. Unfortunately in recent years those colorful and descriptive call signs have gone by the wayside in favor of the nondescript, but more secure, alphanumeric ones generated by the NSA's computers. After all, "Echo-Five-Bravo" just doesn't sound as sexy as "Grim Reaper."

One call sign that sticks in my mind is the one used by Staff Sergeant Jimmie Howard during the battle which would later come to be known as "Howard's Hill." His is a story of tremendous courage against overwhelming odds - there were eighteen recon Marines holding off an NVA battalion - but to me the one incongruous detail was Howard's call sign of "Carnival Time." I wonder who chose that one? I guess it just goes to show the label doesn't matter if you are kicking butt. Howard received the Medal of Honor, and every single one of his men was decorated for bravery. Carnival Time indeed!

One place where the old style call signs are still used is the MSG program. Out there at the embassies we never had access to all of the classified, computer generated stuff, so we just winged it. I did believe in changing call signs periodically for security reasons though, and liked to use a theme when choosing them. For instance at one time the embassy itself was called "Camelot," and each of us took on the name of a Knight of the Round Table. Naturally I, as Detachment Commander, was "King Arthur."

After a while I discovered there was another organization that liked to use handles similar to ours, but for a different reason. Secret Service details traveling abroad with VIPs would "name" those they were protecting, and sometimes the designations were quite comical. A good example is the one given to then-First Lady Hillary Clinton while she and the President were on an official visit to London. Keep in mind these names were not normally released to the public, but occasionally there was a "leak." During that time a good friend of mine was with the Marine detachment in London, and part of his job was to coordinate with the Secret Service detail protecting the first family. In doing so he learned a few choice bits of information - including Hillary's call sign. She was known as "Thunder Thighs" - and if you have ever seen that woman in a dress it's pretty obvious why.

But that wasn't the last call sign she would have. In 2003 now-Senator Clinton visited Afghanistan, ostensibly on a "fact finding" tour. The troops who were there knew it was really a political trip designed to build her résumé and get some press, and naturally they felt the whole dog and pony show was an unnecessary distraction for them. So when it came time to choose a designation for the helicopter she would be riding in someone came up with a call sign that was both accurate and poetic - *Broomstick One!*

IT'S A SMALL CORPS

"It is better to die on your feet than live on your knees."
– Dolores Ibarruri

Sometimes I am truly amazed at how small and closely knit the Marine Corps community is - and one story in particular really brings that home for me. My travels in the Corps have taken me from my original home in New York to the far corners of the earth and finally to my current home in Florida, but unbeknownst to me there was a thread binding all of those travels and experiences together - much like a game of "six degrees of separation."

Early in my career I spent some time in the reserves up in New York, and many of the reservists I served with were New York City police officers and firefighters. I didn't know it then, but many of the police officers I knew there would one day end up working on the NYPD with an officer named Mike Curtin, and Curtin would eventually become a Marine reservist himself after leaving active duty. Readers of my book *Swift, Silent and Surrounded* will recall that it contained two stories about him - the first was about how he helped excavate a Marine's body after the Oklahoma City bombing, and the second told of his own death in the World Trade Center on September 11.

Throughout my career I spent a lot of time serving in the Marine Corps' reconnaissance community, which is very small and close knit. Time and again I would cross paths with the same Marines, and after awhile thought of many of them as family. Some of those very guys were riding in a helicopter which crashed a week before my retirement from the Corps, and their deaths were what inspired me to write

Swift, Silent and Surrounded. Two of those Marines were named Vince Sabastaenski and Jeff Starling.

Once I had retired and moved to Tampa, Florida this story followed me there as well. About a year after I arrived there I received a call from Captain Warren Dickey, who had been my OIC at Camp Pendleton.He had been assigned to MacDill AFB, and soon ended up buying a house half a block from my own. The reason I mention this is he was the one who had told me about the crash in California, and had escorted Vince Sabastaenski's body back to his home in Maine.

The connection to Starling was even more personal. He was a communicator by trade, and I had personally seen to it that he came to join us at 1st Force Recon when he rotated back from Okinawa. He was a solid Marine and a good communicator, and I considered his recruitment to be something of a coup. I didn't know at the time, but he was from Daytona Beach, which is only a two hour drive from my eventual home in Florida.

After I published *Swift, Silent and Surrounded* I came into contact with Jeff's parents, Grandle and Charlotte. As it turns out Grandle is himself a former Marine, and after he read the book we began communicating via e-mail and quickly developed a friendship. Little did I know it, but I would soon meet them as well. That occurred on the fourth anniversary of their son's death. They drove over from Daytona to place flowers at the SOC Memorial at MacDill AFB, and I joined them there to pay my respects.

It all came full circle when I received an e-mail from a former Marine named "Red Bob" Loring, who was interested in reviewing *Swift, Silent and Surrounded* for Leatherneck magazine. He told me he hadn't read the book yet, but had heard good things through the grapevine. I agreed to drive up to Bob's home to meet with him, and when he greeted me at the door I got quite a surprise. Bob was wearing a t-shirt bearing the name of his Marine Corps League Detachment in Zephyrhills. It was a relatively new detachment, and had been named after a Sergeant Major named - you guessed it - Mike Curtin!

LETTER FROM KHE SAHN

By T. M. Barmmer

"Winners never quit, and quitters never win."

During wars, distorted reports are often made from the front lines by people with an agenda. For example, the media dwells on the negative occurrences in Iraq as we work to restore order there and rebuild that country, and pays scant attention to all the progress which has been made. And then there are the allegations made by people like Senator John Kerry who, upon his return from Vietnam, accused our troops of regularly committing atrocities against the Vietnamese people. If what he said was true I wonder why he, in his position as a Naval officer, took no steps to prevent or report such conduct while he was there. This letter from a Marine at Khe Sahn shows the other side of the coin:

Dear People of the News,

The "Folks Back Home" sent me a copy of the *Westport News,* which I appreciate very much, and on the front page of a recent issue you printed a letter from a soldier in Vietnam. Now I ask you to print a letter from a Marine in Vietnam - rather, ten Marines, for it was my squad which read the letter and encouraged me to send one.

My job, and that of my squad, is to search out and destroy the enemy by operations as small Recon patrols, and as large Battalion sweeps.

One soldier saw an accidental shelling of a village - he said it was accidental. And from that one mistake, a few people draw drastic conclusions as to the method of "war" over here.

We are not interpreters - we are Marines. When we got here we fought from Da Nang to the DMZ and we have seen killing, both necessary and unnecessary, on both sides. But one thing those few people don't understand is that there are some people who are willing to die fighting for a freedom that those same few were born into.

I wonder if that soldier has ever seen what will happen to a village if the people don't do exactly as their enemy dictates, such as giving them women and the majority of their crops. We have seen tortures and depravities unfit to print. Yet a few want us all to leave and let these people strive for themselves!

The only people who don't, the only people who can't, appreciate freedom are those who have been given it free of charge. Those very same few. War is hell - but sometimes war is necessary to uproot an evil too deep to scrape off.

There was such a war in Heaven even before earth was created. And war will end only when man no longer exists in his present form. I wonder if that soldier has ever seen us guard a village during rice picking, so the people can work without the fear of death they've grown so accustomed to.

Has he ever seen Seabees build a decent building to house their crops, to keep them from rotting during the monsoons? Has he ever seen Corpsmen risk their lives to stay in a village to help the sick and old, friend and enemy alike? Has he ever seen a Marine give a little girl a piece of candy, or carry a little boy across a stream and teach him to use common soap? We have, because we've done it ourselves.

No part of any war is pleasant - but we ask one thing - why shouldn't we stay and help people who want to be helped, and give them a chance to live the life we take for granted? Why do accidents such as he related become distorted and magnified to represent all conditions here?

If anyone can answer us, we would appreciate one. And we leave you with a little to think about. As long as there is one person left in Vietnam who needs and asks our help to achieve a way of life suitable of living, we will stay, and we will fight.

When America fought for her independence and freedom, the few countries that came to her aid were never accused of such "atrocities of war." We know what freedom is worth, and will go to almost any means to help a country such as this one achieve that which they want and need to survive.

You showed courage by printing the letter from a soldier. We ask you to show the same courage by printing the letter of a Marine - or rather, ten Marines.

Gratefully yours,
L/Cpl T.M. Barmmer
and the 2nd Squad, 3rd Platoon
3rd Recon Bn
3rd Marine Division

NOTE: Lance Corporal Barmmer was KIA on Jan 30, 1968

KHE SAHN REVISITED

By Lee Webber

"The eyes of the nation and the eyes of the entire world, the eyes of history itself, are on that brave little band of defenders who hold the pass at Khe Sahn."
- President Lyndon Johnson

It is all so far behind me, but the vivid memories make it seem like only yesterday. For many years I've talked with people about my experiences during my time in Vietnam. Some of those experiences were full of joy, while many others were filled with pain.

Now on the thirtieth anniversary of the siege at Khe Sanh the time had come for me to return and face the people and the land with whom I did battle so many years ago - the people who fought so tenaciously for their homeland, and the land which was always there. The lessons learned from Vietnam turned me into a tough-minded, resilient survivor. But it also brought compassion and an acutely sensitive form of human understanding into my life. I have come back to Vietnam to face the ghosts of my past. Those same ghosts that have come together with the balance of my life experience to form the man I am today.

I recall our landing in DaNang, looking at the countryside as we descended, wondering where the enemy was - not knowing the real question was: Who was the enemy? As the plane door opened, the hot muggy air of Vietnam filled the cabin with an unforgettable pungent scent that became part of my life as a Navy Corpsman for the next twelve months. During that first night I met a Marine who was heading home and asked him for advice because he had obviously survived his tour. He told me to volunteer for Recon duty. I did, and was sent to Phu Bai and attached to Delta Co, 3rd Recon Battalion, 3rd Marine Division. At Phu Bai we ran eight-man patrols from September to December 1967, then we were moved north to Quang Tri.

Then on January 19, 1968 our platoon was selected to move to Khe Sanh combat base in an effort to reinforce the Recon unit there. Unbeknownst to us, we were moving from the frying pan into the fire. We were to become center-field players in U.S. military history, because on January 21st the base came under siege. As the first rounds landed we scrambled into the makeshift trenches that we had been digging so half heartedly. We huddled together and crouched as low as possible in the shallow trenches we'd dug.

In the days that followed, I rummaged through what was left of our battalion aid station trying to salvage medicine and supplies. I could see planes bringing in what I thought were reinforcements. Instead, down the street came people with cameras and tape recorders shooting pictures, video, and asking the proverbial stupid question, "How are you doing?" I recall looking at one reporter (whose cameraman had poked his lens in my face) and telling him something like, "We need guns, and they send us you guys with cameras…"

As the days wore on they blew our ammunition dump, and we battled the nightly procession of rats as they searched for food and safe haven from the battle. The North Vietnamese then cut off our only water supply, creating the need to air drop water. Near the end of the siege I was evacuated to DaNang hospital for an appendectomy, and then on to the *USS Sanctuary*. When I returned to my unit, the siege had ended and we were located at Dong Ha.

Returning to Vietnam was always a desire of mine. At the quadrennial reunions of our Marine unit there were always discussions that centered on the idea. Two years ago fellow team member Ken "Bernie" Burnett and I decided we were going back. We found Military Historical Tours out of Arlington, VA and began our plans to make our return on the thirtieth anniversary.

The group of more than fifty departed from New York, Los Angeles, Singapore and Guam and met in Hong Kong for the final leg of our journey to Hanoi. We arrived in Hanoi late in the afternoon, and were treated to dinner by the Marine Security Detachment of the America Embassy. It was a good evening of laughter, photos and fun. On day two, as the group moved through Hanoi, I went to Ninh Binh province for the dedication of the Hong Phong children's school. The school, which was paid for through Gannett Foundation grants, was built for pre-schoolers. It was a good feeling to be around the children listening to their singing, laughter and paying attention to their endless curious comparisons between their size and mine. The dedication ceremony was a very appropriate way to begin my return, given the circumstances that existed when I left the country in 1968.

The next day we went to the Ho Chi Minh mausoleum. While I didn't tour the mausoleum, I found the endless line of thousands of pilgrims from the countryside an odd curiosity as they stood quietly, without hats, and hands dangling at their sides. Foreign visitors were lined up, given strict conduct instructions, and lead to the front of the line ahead of the Vietnamese who had been waiting so patiently.

Later that morning we moved on to DaNang. My landing there reminded me of my arrival more than thirty years before. The door to the plane opened, the muggy air filled the cabin, and the only thing missing was the smell of jet fuel.

We then boarded buses and headed to Route 1 for the trip to Hue. Along the way we made stops at Red Beach, Hai Ban Pass and Lan Co fishing village. The countryside has not changed as people continue to live off the land on an almost day-to-day basis. The most obvious changes were the sporadic cellular phones, endless karaoke bars and the more colorful, slightly western-style dress of the people. We arrived in Hue around five in the afternoon to a beautiful sunset dinner on the Perfume River - a far cry from my last meal in the city which consisted of c-rations and water eaten along a dirt road.

The next morning we visited the Cathedral where we met Father Joseph Phuc, who was also at the church in 1967. At the time he had brought his children's choir to perform for the troops, and as he lead us in prayer the room fell silent while men wept. You could almost hear our tears hitting the cold tile floor. At lunch that day we happened to meet Judith Hansen, a former Red Cross volunteer who worked in DaNang from March 1966 through October 1967. She was as surprised to find us there as we were to see her.

We all rose, stood at attention, and sang the Marines' Hymn for her as she stood proudly at the front of the restaurant. As I looked around the room I wondered what thoughts were passing through the minds of the Vietnamese staff who were watching. A couple days later we were in Lang Vei, where we spent hours walking through what was left of the base that had been overrun by the NVA in 1968.

The area was littered with old munitions and rounds. One of the cameramen following us stepped on a mortar. We told him to stop as he rolled his foot forward and backward - we all cringed in hopes it would not explode. Had it exploded, it would have been a horrible way to end our tour and possibly the life of this young cameraman. After lunch we moved on to Khe Sanh where we were spent the afternoon wandering the combat base that was once our home. Bernie and I found what we believed to be the location of our old bunker, Charlie Med, and the Ammo dump.

The base, bulldozed by the Americans following the end of the siege, has since been turned into what can best be described as a disjointed coffee plantation. Earlier, Khe Sanh residents tried to find locations of an old base based on the photos we had. The most prominent location was hill 1015 directly across the valley from our old bunker site. As the clouds and mist rolled away in late afternoon we were able to see the mountain and surrounding hills - the same view we'd had while under siege thirty years earlier. We brought a large wreath and conducted a ceremony as General Carl Mundy spoke of valor, heroism and the bond of brotherhood shared by those who had fought on this hallowed ground. There was not a dry eye while the Marines' Hymn, Navy Hymn, and Taps played. One obvious thing during the entire tour was the friendliness of the Vietnamese people. It was a good feeling to come back to such a warm welcome given the circumstances under which we had left.

At each location roads had been changed and most American structures had been destroyed. At most of old bases there were monuments noting the location and the fact that "the Imperialist American forces and their Saigon puppet regime" had been defeated at that site.

The next day we headed for the Vinh Moc tunnel complex. These tunnels were constructed in 1965-66 by the people of Vinh Linh village. They consist of sixty-eight kilometers of tunnels that go down three levels, the deepest of which is twenty-three meters below the surface. There are thirteen entrances (seven facing the ocean, and six facing the hills). They contain living facilities including a birthing room where several children were born.

The following day we visited the old Dong Ha combat base that is now a residential area containing only one of the original buildings. This appears to be an old theater that, while horribly run down and dirty, is still in use today for videos and laundry. Then it was on to Quang Tri airstrip. We couldn't find the base where I'd lived, but I was able to approximate our old location. Again, the purging of any American remnant of the war was very obvious.

At Phu Bai we found the old air control tower and building slightly renovated but essentially intact. It's one of the few American structures still standing. While in DaNang I visited the Ho Chi Minh war museum containing historical items from 1925 and through the American occupation. There was a photo showing pieces of B-52s noting that "this was the four thousandth B-52 shot down" since the beginning of the war. Others noted the number of Americans killed with the particular weapon on display. We spent an afternoon at the new China Beach resort and Marble Mountain areas, which were much different than I remember.

It is ironic how pictures in our minds can affect our lives and the lives of those around us. This trip back to Vietnam was filled with those pictures, as well as memories of smells, tastes and sounds of the past. Change was extremely obvious everywhere we looked. I believe many were expecting to see things as they were when they'd left Vietnam. There were looks of loss on many faces as the endless searches were conducted for those places in our minds where we hide the pain of the past. When memory was reconciled with today's reality, there seemed to be a release of these ghosts. They were replaced by tourists, new roads, different buildings and scenes of rice fields and playing children. While some of this was a relief, there also was a sense of loss.

It was good to see Vietnam slowly growing and improving. In spite of the damage inflicted upon the people and the land by that horrible war, nature had seen fit to replace the destruction with a new coat of paint. The people were friendly, spoke of growth and peace, and how important both were to them. However, beneath the surface memories of the war lingered. At the museums, displays recounted the atrocities inflicted upon the people by the Americans and the puppet Saigon government. These museums were always filled with guides taking children and tourists around explaining the so-called war crimes. It was interesting to see the looks we would get when these tour groups passed us by.

I am now home with mixed emotions, after having made this journey with other warriors - laughing, crying and sharing our lives together as we did so many years ago. It was a very moving experience, but one that carried with it a much different, maybe older and wiser, view of the land and the people. I for one am walking away with a better, kinder view of this country and her people.

A VISIT TO ARLINGTON

"All who now wear, or have ever worn, the Eagle, Globe and Anchor share a common bond." - Senator and former Marine Charles Robb

The following letter was received by Marine Barracks, 8th&I. The Commanding Officer passed it along to his troops with the following comment: "'We' are different. And I appreciate each of you who have given this dignity and 'love' to our fallen Marines. Semper Fi!"

On Friday, April 28th my family - myself and my brother's family visiting from California - were visiting Arlington Cemetery. We saw the usual sights; Lee's home, the Kennedy gravesite, and the Tomb of the Unknown Soldier. As we were preparing to depart, far off in a treeline, I caught the sight of Marine Dress Blues. Two platoons of your Marines, accompanied by a contingent from the Drum & Bugle Corps were practicing drill. I couldn't leave without finding out what was going on, so I stopped a Major who told me they were preparing to conduct a funeral service for a WWII Marine veteran. Never one to pass up an opportunity to observe Marines in action, I "modified" our sightseeing plans so we could observe the preparations, including the Caisson Platoon from the 3rd Infantry. That decision to change our plans turned out to be one of the best calls I ever made.

While we did not intrude on the actual funeral, we were able to observe the dignity and professionalism exhibited by all the Marines as they formed up, greeted the family, transferred the urn to the flag draped casket and slowly marched off into the cloudy morning as the band struck up that familiar old hymn "Onward Christian Soldiers." It was a truly moving experience. One that reinforced my pride in our Corps, and our continuing commitment to care for our own.

As the procession moved off down the street I was asked by a passerby and his family why the Marines were there. I told him they were burying a Marine from WWII. He shook his head and said, "That's why you guys are who you are. You care about each other. I'm an Air Force vet, and I'll bet we don't do much of anything at all."

Most meaningful to me was the reaction of one of my sons who has grown up as a Marine kid but never demonstrated too much interest. He sat off by himself on the grass watching the whole event, never saying a word. As we left the Cemetery that morning he sidled up to me and told me that Arlington was the best place we had visited, and that he hopes to be a Marine. I thank your Marines for that.

That chance morning encounter in Arlington moved my entire family, increased the bond between my boy and myself, and provided a reminder to me of how wonderful it is to be in our Corps. Please accept my profoundest thanks for that unexpected opportunity.

WHY ARE WE MARINES?

By Patrick A. Rogers

"'I must do something' will solve a lot more problems than 'something must be done.'"

A long time ago I went into a local deli for a sandwich. The elderly counter guy looked at my haircut and stated "What are you, in 'da army or sumpthin'?"

"No Sir," I stated, "I am a Marine."

"Same shit," says he, his frame of reference obviously having been poisoned over the years. As my late dad used to say, "Never enter into a battle of wits with someone who is unarmed." The store lost a sale and a customer, but I kept that exchange in the data banks for years.

Fast forward to June of 2003. I was at the Activation Ceremony for the Marine Corps Special Operations Command Detachment 1, something long overdue for us. It had been a long and frightful trail leading up to the activation, and the celebration afterwards was noteworthy. Especially when Dick Torykian, a former Marine Officer from RVN days and currently a driving force in the Marine Corps Law Enforcement Association, addressed the dinner attendees. Dick is not shy, and his comments are usually hilarious. We were slightly taken aback when he said that with all of the luminaries present (including one former CMC) he would recognize only one - a Navy two star. The Admiral acknowledged the honor, and Dick commented, "After all, the Marine Corps is part of the Department of the Navy." After a pause (which could have been a pregnant pause, but the gestation period was closer to pachyderms than to humans) Dick stated, loudly and focused "The *Men's* Department!"

When the phone rang at 0600 the following morning, I knew that it couldn't be good news. It wasn't. It was my girlfriend's brother making the notification of the death of her father. Cornelius McCarthy was an Army veteran of World War Two, and as we made travel arrangements for the trip to Florida I commented that Ellen's Mom should contact the Army for Funeral Honors - but when we arrived at her family's home Mrs. McCarthy told us that she couldn't find any discharge papers, without which the Army would not provide a Funeral Detail.

While I understood their desire not to chance providing a Detail for a "poser," I also understood that many families would unfortunately not have access to certain paperwork. Mrs. McCarthy did have her husband's VA Claim Card - "He had been bayoneted," she told us - but this was apparently insufficient.

I was irked. A veteran has earned the right to be recognized for his service to his country, and no greater honor can be rendered than to have fellow warriors present for his final journey. I called up the Commanding Officer of the Det, Colonel Bob Coates. There is no finer warrior or leader in the world, and I figured he would have an answer for this dilemma.

Not very long afterwards the cell phone rang, and the voice on the other end said, "Sir, this is First Sergeant DeWitt. How can I help you?" Tom DeWitt is a former Force Reconnaissance Marine, and had been the Platoon Sergeant when a CH-46 crashed into the sea during a VBSS on the USNS Pecos in December 1999. He was a strong leader, and perhaps best represented what a Platoon Sergeant should be. He was now assigned to the I&I Staff at 8th Tank

Battalion in Tallahassee, which was a good seven hour drive away. After I explained our problem he said simply, "We'll be there," and on the day of the service he and four other Marines from the I&I Staff arrived at the funeral home and executed perfect honors for a fallen warrior.

Major Cornelius McCarthy had enlisted in the Army in December, 1940, at age twenty-four. He served initially in the Aleutians, but was commissioned and reassigned to the 90th Infantry Division. He landed at Normandy, and fought through into Czechoslovakia. After being wounded in France he was evacuated to the UK but eventually rejoined his command, and was wounded again on May the 2nd,1945 - one week before the end of the European war.

During his tour in Europe Major McCarthy was shot twice and bayoneted once, and received three real Bronze Stars, two Purple Hearts, and the most coveted of all combat awards, the Combat Infantry Badge. He was, truly, a stud. Like many of his generation he returned home and spoke little of his time in the War - but it was apparently still on his mind. He kept his .45 under his mattress, and a 1:50,000 map of Bavaria- appropriately annotated and close by.

And so First Sergeant DeWitt knelt in front of Mrs. McCarthy and said, "On behalf of the Secretary of the Army and a grateful nation, we offer this flag as a token of your husband's service to his country," and thus one warrior - a soldier, was honored by other warriors - Marines.

There was no hesitation on the part of Tom DeWitt. A Marine needed assistance. No documentation was necessary. No mention of the fact that they drove for seven hours, rehearsed and then rendered Honors, and then had to drive seven hours back to station. They weren't too busy, too tired, or too unwilling.

They did it because they are Marines. Part of a breed whose ethics represent something that is unfathomable to the socialists, apologists and wimps that seem to be most vocal in our society. We are different. A former Assistant Secretary of the Army under the administration of a Commander in Chief who was impeached felt that Marines were radical.

Of course, I have always wondered why we hold Lance Corporals to a much higher standard of conduct than the President of this Nation. Then again, a Lance Corporal is a Marine, and that President was just a worm. The Marine is the better man. And that is why I am proud to be a Marine.

Patrick Rogers is a Vietnam veteran, retired NYPD Detective, and retired Marine Corps Reserve Warrant Officer.

A MARINE COMES HOME

Dorothy Rabinowitz

"Cowards die many times before their deaths; The valiant never taste death but once." - William Shakespeare

Fallen warriors remind us why whiny celebrities are irrelevant. The battle of Iraq may be over, but the warriors for peace struggle on. Theirs is not an easy road, particularly in the entertainment industry, which is packed with notables fresh from their vocal campaign against the war, the president, Donald Rumsfeld and Dick Cheney - objects of scorn in all the best circles, from Paris to California.

Now, it appears, some celebrities worry about damage to their careers. The Dixie Chicks have taken a hit. Sean Penn thinks his views have cost him jobs. Tina Brown, whose main concern about the war seems to be that it caused the postponement of her new TV show, announced last week that it would soon air and that she planned to decorate the set with an American flag bigger than anyone else's. She had to scrape up as many core American values as she could, declared Ms. Brown, "to have any hope of being allowed on TV at all in the current climate of punitive patriotism."

No fear. Americans aren't likely to concern themselves much with Ms. Brown's flag, in the event they actually encounter her program. Most of them have matters more pressing on their minds. For some, these days, those matters include funerals and mourning rites for people they have never met.

On April 14 in Vermont, for example, mourners gathered for the funeral of twenty-one-year-old Marine Corporal Mark

Evnin, killed in action on the drive to Baghdad. A thousand people attended the rites at Ohavi Zedek Synagogue in Burlington, at which the Marine's grandfather, a rabbi, presided. Reporters related how the Marine Corps League color guard and local firefighters flanked the walkway into the synagogue, where mourners included the Roman Catholic bishop and the governor.

Crowds lined the streets in salute - some with flags, some with signs - everywhere the funeral procession passed. But what struck the *Burlington Free Press* reporters most were all the strangers who had been impelled to come to the cemetery to honor the young Marine. One of them was a mother who had brought her two young children and stood holding two American flags. "Every single man and woman out there is my son and daughter," she told the journalists. "He could have done a lot with his life. But he gave it to the nation."

Two days later came the funeral mass for Marine First Lieutenant Brian McPhillips of Pembroke, Massachusetts, who had been killed not far from Baghdad. Three Marines died in the firefight at Tuwayhah described by *Dallas Morning News* embedded reporter Jim Landers. The 2nd Tank Battalion had run into an ambush by a band of Islamic Jihad volunteers - Syrians, Egyptians, Yemenis and others. Lieutenant McPhillips went down firing his machine gun.

The knock that brought the news home in the early hours of April 6 had caused the walls to reverberate, his mother recalled. His father, a Marine veteran of Vietnam, knew at once what the five AM visit meant. They never come because somebody's been wounded: "They want you to know as soon as possible."

Neither of the McPhillips was surprised at Brian's choice of a military career. His father had served, and his great-uncle had fought at Guadalcanal. Julie and David McPhillips had been the sort of parents who wanted to imbue their children with a consciousness of history - that of their country's at least. So they took them to places like Shiloh, Antietam, Gettysburg and other national shrines.

David McPhillips nevertheless used all his powers of persuasion to keep Brian from enlisting in the Marines right out of high school. Heeding his parents, Brian went off to Providence College, a Catholic institution, where he thrived, compiled an academic record most people considered enviable, his father included, and looked to the future. Shortly after graduation in 2000 it arrived, with the commissioning ceremony that made him an officer in the Marines. He would go to war, his father reported, carrying his rosary and his Bible.

At his funeral service at the Holy Family Church in Rockland, where Brian's mother attended daily Mass, David McPhillips recalled his son's generosity and enterprise. Mrs. McPhillips would deliver a eulogy of her own, afterward carried in the local papers, on the subject of her son's life and death. She saw herself, Julie McPhillips said, as one of the fellow Americans for whom he had given his life. It had been her great privilege to be his mother: "To you my dear and faithful son, from earth to heaven I salute you . . . ."

As at Corporal Evnin's funeral, crowds lined the streets. Brian's uncle Paul Finegan pondered the problems getting to the cemetery in Concord - a 150-car cortege traveling fifty miles on the busiest highway in New England. He had, it turned out, nothing to fear. Fifty state troopers, many of them coming in from days off, had closed most of the road for them, a stretch of thirty-five miles.

Then came another sight he could scarcely believe. At the side of the road, near their halted cars, stood streams of people, standing at attention - paying their respects.

"They stopped all these cars, and people got out to stand holding their hands over their hearts," he marveled.

He should not have been surprised. Scenes like this are the reason all the celebrity protesters can stop worrying about public wrath and punishment. Americans have other things on their minds all right. September 11, for one. What they have on their minds, too, since the just-concluded war, is the consciousness of who they are and what this society is that it should have produced men and women of the kind who fought in that war and died in it.

People got a powerfully close look at their fellow Americans in uniform these last weeks. This is what impels them now to stand at roadsides in tribute, heedless of where else they had to go. And this is why strangers flock to funerals.

From Dorothy Rabinowitz' Media Log

GENDER, LIES AND VALOR

By LtCol Matthew Dodd

"Heroes are people who rise to the occasion and quietly slip away." - Tom Brokaw

I recently learned that the Army has seen fit to award the Bronze Star Medal to PFC Jessica Lynch, and while I am not surprised, I am truly offended. The initial press reports following her capture claimed that she had fired her weapon until she ran out of ammo and killed several Iraqis, but that piece of fiction was obviously intended to further the cause of those who advocate the idea of "women in combat." In a television interview Lynch revealed that she did not fire a shot, was unable to even load a magazine into her M-16 (probably due to poor maintenance on her part), and that her only contribution to the ensuing firefight was to kneel down in the vehicle in which she was traveling and pray for salvation.

My own Father served in the Army, and received the very same award as Lynch - but he got it for knocking out a German machine gun nest during the Battle of the Bulge. I feel his decoration, along with those of countless others, has been tarnished by the actions of the clueless individual or individuals who engineered this "award." I hoped these sentiments of outrage and indignation found their way into PFC Lynch's hands, and if she had any notion of what is right she will have publicly refuse to wear the award and in doing so honor the true heroes. But that didn't happen. After all, she was just a pawn in a much bigger game. The feminists in this country are so desperate for a standard bearer for their agenda of military "equality" that they decided to invent a "hero" with the aid of the liberal media. Now they can point to the "most famous hero" of the Iraq War and say "I told you so." What is truly sad is everyone in this country knows the name Jessica Lynch, but only a handful know about my old friend "Horsehead" Ed Smith and the others who died in actual combat. I guess that's because their deaths didn't further someone's political ambitions and are therefore insignificant. The only way to end this nonsense is to give the feminists what they claim to want - 100% integration. And that includes universal draft registration. Then land the unwilling female draftees on a hostile shore under fire and let the silent majority of American women who don't want any part of combat put the feminists out of business!

But enough of what I think. Lieutenant Colonel Matt Dodd presented the facts, and they speak for themselves:

I, like most Americans, was very relieved and happy when I heard the news that PFC Jessica Lynch and her fellow soldiers were rescued. After hearing the horror stories about the torturous methods of the former Iraqi regime, it was all too easy to imagine the unimaginable happening to our soldiers being held as prisoners of war. Knowing they were all back under coalition control was joyous news.

The public media coverage of our Iraqi POW saga was impossible to ignore. I, too, was glued to the television when PFC Lynch arrived home and addressed the nation. I found her to be a charming, humble, sincere young woman who was obviously proud to be a soldier and extremely thankful for all the love, support, and encouragement she and her family received all throughout her ordeal. I am grateful that she is alive, and for the fact that we have such fine young people voluntarily serving in our military.

However, despite all the hype, emotional appeal, and the apparent dedicated lobbying efforts of many pushing their own agendas, I am vehemently opposed to attaching the label “hero” to PFC Lynch.

PFC Lynch was a POW, not a wartime hero. A dictionary defines hero as “any person admired for courage, nobility, etc.” From all that I have been able to learn about the circumstances surrounding her capture and her actions leading up to her rescue, I have seen nothing noble or courageous to admire.

She was a passenger in a vehicle in a convoy that took a wrong turn and ran into a deliberate ambush. Her vehicle crashed and she was so severely injured in the crash that she was knocked unconscious and unable to fight or resist capture. She was taken prisoner by her ambushers, given excellent medical care in a hospital, and was subsequently rescued from that hospital in a well-executed raid by well-trained forces.

From what I just summarized, the U.S. Army decided to give her a Bronze Star Medal with the following citation excerpts:

“For exemplary courage under fire during combat operations (from Mar 23-Apr 2, 2003)… Private First Class Lynch’s bravery and heart persevered while surviving in the ambush and captivity…. [Her] belief in [her] Battalion’s motto “One Team, One Fight” is in keeping with the finest traditions of military service. Her honor, courage and dedication reflect great credit upon herself, the 507th Maintenance Company, 3d Infantry Division, Victory Corps, and the United States Army”

I have never written an Army Bronze Star Medal recommendation package. However, I have seen and written many Marine Corps Meritorious Masts (authorized and awarded by company commanders) for deserving Marines that contained more truth, details, and substance than PFC Lynch's pitifully weak citation above.

I suspect that Lynch's award was pre-approved at the highest levels, and that the task of writing the award package became a mere administrative "check-in-the-block." I cannot believe that her award package was initiated and submitted by her unit, and subsequently judged on its own merits against other submitted awards, and then approved all the way up the chain of command by her Corps Commander and the Secretary of the Army. Some news reports allege that Army officials pressed for a Silver Star medal for Lynch, but yielded when her unit resisted.

Regardless of how and when the decision to award the Bronze Star to PFC Lynch was made, it proves to me that the Army today has a blatant, systemic disregard for maintaining the highest standards for its highest combat awards.

Let me share with you my analysis of Lynch's citation as if I were a member of an awards board somewhere in PFC Lynch's chain-of-command:

Since PFC Lynch was either unconscious or incapacitated following the vehicle crash, her "exemplary courage under fire during combat operations" was limited to the few moments between the initiation of the ambush and the vehicle crash. I understand she did not fire her weapon at all, so I wonder how she demonstrated her courage under fire.

Next, her "bravery and heart persevered while surviving in the ambush and captivity." That sentence tells me that she did not give up her will to live despite her extensive injuries. Being a fan of individual character, I appreciate her choice, but I do not see that character trait being justification for a combat award.

Her battalion had a catchy motto. She apparently believed in that motto. How did she show her belief in that motto, and how does her belief in that motto live up to "the finest traditions of military service?" Was she special among the members of her battalion by actually believing in the battalion motto?

As far as her honor, courage, and dedication reflecting great credit upon herself and her entire chain-of-command, I just do not see any evidence that she did anything above and beyond surviving her horrendous injuries and not giving up her will to live. I would hope most of her fellow soldiers did or would have done the same exact things if they found themselves in the same circumstances.

I have absolutely nothing against PFC Lynch. My only complaint is with the leaders in the chain of command who approved and then thrust this combat award upon her and nurtured the false myth that she is a heroic woman warrior. She was simply a good soldier who survived a tragic, combat-related accident of incompetence and was rescued by warriors who did more to earn the label "hero" than she did. Where are the awards and public media coverage for those heroes who rescued Lynch?

I admit that I am biased in my assessment of POW Lynch as a mislabeled war hero. I am biased by the heroic citations of our former POWs who earned the Medal of Honor. Two examples in particular:

Air Force Major George E. "Bud" Day suffered a broken arm in three places and a badly injured knee when he was shot down in North Vietnam. He was captured, and interrogated and tortured in a prison camp. He escaped and was eventually ambushed, wounded again, re-captured, and returned to his captors. His citation noted his continuous maximum resistance as "significant in saving the lives of fellow aviators who were still flying against the enemy."

Navy Captain James B. Stockdale ejected from his crippled plane and parachuted into North Vietnam where he was beaten in the streets by an angry mob, bound and captured, and refused favors in exchange for medical treatment on his severely broken leg. Recognized as the senior ranking U.S. prisoner responsible for organizing widespread resistance to their enemy captors, he was singled out for interrogation and attendant torture. Using self-disfiguration and inflicting a near fatal wound to himself as symbols of his willingness to die rather than capitulate, his actions led to his captors easing up on the harassment and torture of all prisoners, and "earned the everlasting gratitude of his fellow prisoners and of his country."

Let's put these three citations in perspective. We have two men who were badly injured prior to their capture, brutally tortured over a period of many years, continuously resisted their captors' efforts, and through their self-sacrificing leadership examples inspired their fellow prisoners and helped save their fellow prisoners' lives. They both earned our nation's highest combat honor.

Then we have PFC Lynch, who was given the Bronze Star, our nation's fourth-highest combat honor, for being in the wrong place at the wrong time, suffering horrendous vehicle accident injuries that prevented her from resisting capture, receiving life-saving medical attention from her captors, and being rescued in a daring raid about two weeks after she was first captured.

What does the vast disparity in the standards for these combat awards say about the relative value and fairness of our combat awards system? Do we have a double standard for combat awards based on gender expectations?

One former Marine's e-mail to me said it best and inspired me to write this article:

"So let's use Lynch as the foundation for future medals. Since she never fired her weapon, then anyone who does fire it (at the enemy) receives a Silver Star. To shoot at the enemy and be fired at and even hit back, you receive the Distinguished Service Cross. To shoot the enemy and get hit back and actually kill the enemy, wow, you get a Purple Heart, campaign ribbon, combat action ribbon, and Medal of Honor. At the rate this is going, I am going to find out what company is authorized to sell these medals and buy stock."

My hat is off to PFC Lynch, the former Iraqi POW, but not to the farcical "war hero" her shameless chain-of-command would have all of us believe she was.

Matt Dodd is a regular contributor to Military.com

THE PHOTOGRAPH

"When we recall the past, we usually find that it is the simplest things, not the great occasions, that in retrospect gave us the greatest happiness." – Bob Hope

In 'Swift, Silent and Surrounded' I was critical of several celebrities who have conducted themselves in a less than patriotic manner, and without a doubt those comments were well deserved. To the best of my knowledge the only one in that entire group who has served in uniform is Jane Fonda - although as we all know that uniform belonged to the North Vietnamese Army. In all fairness it must be pointed out ALL of Hollywood is not like that, and I would now like to balance the books and talk about some of the luminaries whose support for Americans in uniform have endeared them to the hearts of military members over the years.

The first name that always comes to mind is Bob Hope. Beginning with his first show at March Field, California in 1941, Mr. Hope began a legacy of service that would continue for the next sixty years, and would make his name synonymous with 'entertaining the troops.' His recent passing was felt by us all, as he was as much a part of our Armed Forces as anyone who wore the uniform.

Another legendary figure was Martha Raye, who was better known to the troops as 'Colonel Maggie.' The stories of her visits to Vietnam are legion and could easily fill a book, and if that book has not yet been written perhaps it should be. She deserves to be recognized for the comfort she provided to countless servicemen.

There are of course many, many others who have given of their time and talents, more than could possibly listed here. One story that is truly representative of what these fine patriots mean to those of us who have born arms in our Nation's defense follows, and should serve as a lesson to the ungrateful "stars" in our midst:

My husband Richard never really talked a lot about his time in Vietnam other than he had been shot by a sniper. He did have a rather grainy, 8x10 black & white photo he had taken at a USO show of Ann Margaret with Bob Hope in the background however, and that was one of his treasures.

A few years ago Ann Margaret was doing a book signing at a local bookstore. Richard wanted to see if he could get her to sign the treasured photo, so he arrived at the bookstore at noon for the 7:30 PM signing. When I got there after work the line went all the way around the bookstore, circled the parking lot, and disappeared behind a parking garage.

Before her appearance bookstore employees announced that she would sign only her book, and no memorabilia would be permitted. Richard was disappointed, but wanted to show her the photo and let her know how much those shows meant to lonely troops so far from home.

Ann Margaret came out looking as beautiful as ever and, as second in line, it was soon Richard's turn. He presented the book for her signature, and then took out the photo. When he did there were many shouts from the employees that she would not sign it, and Richard said, "I understand. I just wanted her to see it."

She took one look at the photo, tears welled up in her eyes, and she said, "This is one of my gentlemen from Vietnam, and I most certainly *will* sign his photo. I know what these men did for their country, and I always have time for 'my gentlemen.'" With that, she pulled Richard across the table and planted a big kiss on him.

She then made quite a to-do about the bravery of the young men she met over the years, how much she admired them, and how much she appreciated them. There weren't too many dry eyes among those close enough to hear. She then posed for pictures and acted as if he was the only one there.

Later at dinner, Richard was very quiet. When I asked if he'd like to talk about it, my big strong husband broke down in tears. "That's the first time anyone ever thanked me for my time in the service," he said. Richard, like many others, had come home to people who spat and shouted ugly things at him. That night was a turning point. He walked a little straighter and, for the first time in years, was proud to have been a Vet.

I'll never forget Ann Margaret for her graciousness and how much that small act of kindness meant to my husband. I now make it a point to say "Thank You" to every person I come across who served in our Armed Forces. Freedom does not come cheap, and I am grateful for all those who have served their country.

CONAN THE BARBARIAN

"Dogs make such agreeable friends; they ask no questions, and make no criticisms."

The closest I have ever come to visiting Nirvana was the time I spent living in a place called Talega. Camp Talega is located in the far northern reaches of Camp Pendleton, California and was once the home of 1st Reconnaissance Battalion. It was not much to look at - in fact it was nothing more than a collection of a few old buildings and row after row of rusting Quonset huts - but it was a magical place nonetheless. The best analogy I can make is it was almost like living in a mining camp in the Wild West. If you have ever seen the movie *Heartbreak Ridge* you have gotten a glimpse of Talega. Clint Eastwood filmed it right there in our barracks, and did the bar scenes at a hole-in-the-wall named Carl's out in the town of Vista. It was an awesome place.

During my tenure there 1st Recon was commanded by Lieutenant Colonel Wheeler Baker, who was something of a legend within the reconnaissance community. He believed in training hard, in playing hard, and in trusting his junior leaders to handle the details. The result was an atmosphere wherein the Marines in the command developed into a close knit community, and the camp where we lived took on the feeling of a remote enclave.

Back in those days we had our own little enlisted club, which was located in one of the Quonsets at the back of the camp. It was pretty Spartan, but was a great place to go on weeknights or even on weekends when we didn't want to drive the long road to town or to mainside. The best part for me was that our club was filled with a bunch of steely-eyed recon types - all of the pogues, pukes and pretenders were elsewhere. I guess word about the club eventually filtered out to the town, because eventually a few female "groupies" began to find their way out to our little compound. They obviously weren't steely-eyed warriors, but were welcomed nevertheless - for obvious reasons.

During that period we had a battalion mascot - a black dog named "Conan." Nobody was really sure where he came from, and he didn't really *belong* to anyone. He just roamed around the battalion area like he owned the place. During the day Conan would choose a platoon and follow it through the training day, and in the evenings he would open the door to the club with his nose and join us inside. He would then walk from table to table soliciting the donation of table scraps, and when satisfied would park himself in front of the television set until closing time.

Now keep in mind we really didn't think of Conan as a mascot. He was more like one of us. Not content to follow along on PT runs, that dog would even join in on amphibious ops and swim in the surf zone. At one point someone got the bright idea to fit him with a custom rappelling harness, and before you knew it he was rappel qualified. We even tried to get Conan into Jump School at Fort Benning, since there was apparently a "K-9" course there designed for working dogs - but nothing ever came of that.

Toward the end of my tour in Talega the time of another Marine was winding down as well. Corporal Bill Plant, a member of the Deep Reconnaissance Platoon, was about to get out of the Marine Corps. Bill was an exceptionally nice guy, and was pretty old by recon standards - in fact at thirty-something he was considered a downright Methuselah. He had taken a shine to Conan, and had in recent months assumed responsibility for feeding and looking after our four-legged friend. When it came time to leave for his home in Wyoming I guess Bill couldn't bear to part with his companion, so he packed him up and took him along. It was sad for us to see our old pal go, but we knew it was the best thing for him. Every now and then I still catch myself thinking of our buddy Conan, recon Marine extraordinaire, as he happily runs patrols through the wide open spaces of Wyoming. There are no fleas on you, buddy!

BLACKHAWK DOWN

"Retrieving wounded comrades from the field of fire is a Marine Corps tradition more sacred than life."
– Robert Pisor

A number of articles have been written in recent years about the anonymity with which troops who die in combat or training meet their end, and how that contrasts sharply with the high profile deaths of well known personalities. This segment contains two examples of how that is true, and will hopefully cause more Americans to take the time to learn and remember the names of the brave souls who put their lives on the line on a daily basis - not for money or fame, but so that we can live in peace and prosperity.

When I first read these articles I took a personal interest because the 15th MEU Memorial Fund, which has been the inspiration for my books, supports the families of a few good men whose names are only known to their families and friends - and to those who took the time to read the Dedication in 'Swift, Silent & Surrounded.' Their names are there so that they will never be forgotten:

In January of 2003, on a high Afghan plain seven miles east of Bagram air base, a Blackhawk helicopter went down, killing the entire crew. The four U.S. soldiers who died in the accident, like the seven astronauts who perished a few days later aboard the Space Shuttle Columbia, were volunteers, taking on risks they understood well in service of their country. Beyond their units and their families, their deaths attracted little notice - a paragraph or two in some newspapers, not even that in others.

The tragedy of the space shuttle disaster grips the nation as do few other catastrophes, and for good reasons. Even this many decades into space exploration, astronauts embrace dangers the rest of us can only imagine - but that many of us do imagine, and even dream of. As they fling themselves into orbit and float in the void while trying to tell us what they see and feel, men and women like David M. Brown and Kalpana Chawla and the others who died Saturday become more than role models of discipline and courage and good cheer in cramped circumstances. They come to embody national aspirations of greatness, and human aspirations to reach beyond ourselves.

Yet as we read the biographies of these brave seven, replay their buoyant interviews of recent days, and come to know the grief-stricken but proud surviving spouses and parents, we might spare a moment also for the four who died near Bagram, and the others most of us will never hear about. As their remains were transported to Germany for autopsies on the way home, the victims were identified as Chief Warrant Officer Mark S. O'Steen of Alabama; Chief Warrant Officer Thomas J. Gibbons Tennessee; Sergeant Gregory M. Frampton of California; and Staff Sergeant. Daniel L. Kisling Jr. of Missouri. They joined eighteen other service members who have died in accidents in the Afghanistan campaign and twenty-five killed by hostile fire - a total of forty-seven deaths since the fall of 2001.

Thousands risk their lives every day in that distant country and in the skies over Iraq, and thousands more may soon be asked to do so. With so many reserves being pressed into service and scheduled retirements from the military being delayed, the term “volunteer” is stretched and tested. But these are all people who know, or who knew, they might face danger. These casualties, too, leave empty spaces in the lives of loved ones.

The prayers of a nation were offered in memory of seven astronauts and their families, and rightly so. They gave everything in service to the nation, as did the Bagram four and so many more.

If the Columbia tragedy was an isolated example of the anonymity with which our troops die it would not be so bad, but unfortunately there are many more. One in particular stands out, and because of it I find myself harboring some animosity towards the "rednecks" on the interstate who have a decal with the number "3" affixed to the rear window of their pickup trucks. They just don't get it, and have no true understanding of the word "hero."

On February 18th, 2001, while racing for fame and fortune, Dale Earnhardt died in the last lap of the Daytona 500. It was surely a tragedy for his family, friends and fans. He was forty-nine years old with grown children, one of whom was in the race. I am new to the NASCAR culture, so much of what I know has come from the newspaper and TV. He was a winner and earned everything he had. This included more than forty-one million in winnings and ten times that from endorsements and souvenir sales. He had a beautiful home and his own private jet. He drove the most sophisticated cars allowed, and every part was inspected and replaced as soon as there was any evidence of wear. This is normally fully funded by the car and team sponsors. Today, there is no TV station that does not constantly remind us of his tragic end, and the radio already has a song of tribute to this winning driver. Nothing should be taken away from this man - he was a professional, and the best in his profession. He was in a very dangerous business... but the rewards were great.

But two weeks before Earnhardt's death seven U.S. Army soldiers died in a training accident when two UH-60 Blackhawk helicopters collided during night maneuvers in Hawaii. The soldiers were all in their twenties. They were pilots, crewchiefs and infantrymen.

Most of them lived in sub-standard housing. If you add their actual duty hours (in the field, deployed) they probably earn something close to minimum wage. The aircraft they were in were between fifteen and twenty years old.

Many times parts were not available to keep them in good shape due to funding. They were involved in the extremely dangerous business of flying in the Kuhuku mountains at night. It only gets worse when the weather moves in as it did that night. Most times no one is there with a yellow or red flag to slow things down when it gets critical. Their children were mostly toddlers who will lose all memory of who "Daddy" was as they grow up. They died training to defend our freedom.

I take nothing away from Dale Earnhardt, but ask you to perform this simple test. Ask any of your friends if they know who was the NASCAR driver killed on February 18th, 2001. Then ask them if they can name one of the seven soldiers who died in Hawaii two weeks earlier.

On February 18th, 2001 Dale Earnhardt died driving for fame and glory at the Daytona 500. The nation mourned. Two weeks earlier seven soldiers died training to protect our freedom. No one can remember their names.

This piece was compiled from an article which appeared in the Washington Post on February 3, 2003 and an editorial written by CWO-4 James V. Torney on February 18, 2001

A THOUGHTFUL DEED

By Diane Bell

"What shall I say of the gallantry with which these Marines have fought? I cannot write of their splendid gallantry without tears coming to my eyes." - Major General James Harbord, U.S. Army

For those without relatives in the military, war news can become a blur of daily press briefings and TV news reports. For Teri Merickel, the conflict got up close and personal during a flight from Chicago. She walked aboard her United plane to San Diego behind a Marine captain who was with a young woman. The officer was carrying in his arms what appeared to Merickel to be a beautiful trophy. The two passengers were seated directly across the aisle from her. Merickel admired the "trophy" but didn't have a chance to ask what it was because another passenger quickly came back from the first-class cabin and invited them to come up to that section. After they moved, the passenger returned and took one of the empty seats. He started sobbing.

After a few moments he composed himself, apologized to Merickel and explained. He too was a Marine en route home from Iraq. He informed her that the beautiful "trophy" she had seen was actually an urn containing the remains of a fallen Marine. The wife of the deceased and the urn were being escorted home by the officer.

The story doesn't end there. Merickel soon learned that the fellow who had done this good deed was returning home to San Diego on a brief twenty-six-hour turnaround for the first time in nearly a year.

His nine-year-old daughter had saved all her money to help buy a first-class ticket for her dad. But when he saw the grief-stricken widow and her Marine escort sitting in coach seats, he asked a flight attendant if he could give his seat to the woman, and if the captain could take the empty seat next to it.

When the plane touched down, the pilot announced that a fallen Marine was aboard. Everyone was silent and the passengers remained in their places while the widow and her escort disembarked. As Merickel said goodbye, she asked the Marine passenger next to her if he was going to tell his daughter he gave up his first-class seat.

He thought and then softly replied, “Maybe someday.”

This story appeared in the San Diego Union-Tribune in August 2003

MAYBE YOU CAN BE
One of Us

During the Korean War, Chinese military propagandists told their troops that in order to join the U. S. Marine Corps you had to bring your Mother's head to the recruiter in order to demonstrate your desire to join!

I will never forget my recruiter, a Gunnery Sergeant by the name of Knauf. He was larger than life to me, as recruiters are supposed to be. But the one thing I remember best was him telling me "I don't think you have what it takes to be a Marine," while the other services were offering me the moon to sign on the dotted line. The Marine Corps recruiting office in Huntington Station, N.Y. was at the end of the hall, and in order to reach it I had to pass through a gauntlet of Army, Navy and Air Force recruiters armed with pamphlets showing their technical training, college benefits and opportunities for travel. The Gunny didn't offer me squat, and at times it seemed as if he was interrogating me about why I wanted to join his beloved Corps. In the end it was up to me to convince him that I should be allowed to enlist!

From what I can see things haven't changed all that much. Marine Corps recruiting commercials are all about pride, chivalry, self-sacrifice and overcoming obstacles. The other services have picked up on that to some small degree as the war on terrorism began in the wake of 9/11, but they still rely heavily on the benefits rather than the challenge.

The Corps has had some classic recruiting posters - I can still remember passing one that said "The Marine Corps

Builds Men" outside the local post office when I was barely old enough to walk - but there is one in particular which sums up what it means to choose a Marine's life. You know the one. The leather faced, wild-eyed DI is nose to nose with some hapless first phase recruit, with the brim of his Smokey the Bear cocked over one eye. The words below say it all. "Nobody Promised You a Rose Garden." I know one thing - Gunny Knauf never did!

The following is an excerpt from an e-mail sent from one non-Marine Corps officer to another, and I think it illustrates my point perfectly:

"It's interesting that this should pop up now. I spend my retirement days now in a high school classroom teaching ROTC - not that it has anything to do with this except...

Last week I had a DoD week. That's where all the services plus the Coast Guard come in to my classroom - one each day. The idea is for them to teach high school kids about their branch of the service and what it might have to offer.

I don't mean to start a debate about which service has the best training or "deals," but here are my observations. I thought you might find them interesting.

The Coast Guard - two Petty Officers showed up late for the first class, both wearing class B uniforms. Their pitch consisted of about ten minutes of showing pictures of boats and helicopters and a few minutes of a canned recruiting video - students slept. After that came "this is how much money we can give you" - bonuses - college, etc. They struggled to keep the attention of the students.

The Army - one Staff Sergeant recruiter in class B accompanied by two soldiers from the 101st Avn in Class A's. I thought this would be better. The recruiter showed his canned "Be all you can be" video - students slept - and then got into the "this is how much money I can give you" spiel.

The two 101st soldiers then had their opportunity to sell the Army. The first thing out of the PFC's mouth was how great it was that the Army was paying off his loans, how much more education he was getting, and how he would retire from the Army almost a millionaire because he had been taught about investing his money early - all good things of course.

The Sergeant had much the same speech. The recruiter did add that basic training was tough eighteen years ago when he went through, but it's easy now.

The Navy - one Petty Officer in Class B's. Ditto. Watch the video - student's slept - then the "here's the money" deal.

The Air Force - stood us up.

The Marine Corps - one Sergeant in Marine Corps Dress Blues. His pitch - "I know all the other recruiters have told you about the money they can offer. Every service can give you the same college money deal, and we all get paid from the same bank. An E-5 makes the same money in every branch of the service. That's all I have to say about that. We want you in the Marine Corps because you want to be one of us, not because you're looking for college money."

He talked about Marine Corps Values and how proud he was to be a Marine. He then showed a "no holds barred" video of Marine Corps Boot Camp. It showed Marine Corps Drill Instructors chasing people off the bus (you could see the sweat on the faces of the recruits while they stood on those little yellow feet), in their face - yelling, running the crap out of them, dogging them out on confidence courses and in PT. It showed them the "Crucible" - recruits literally ran into the ground for fifty some hours straight. He then showed the ceremony at the end of the crucible where each Marine that completed it received his or her eagle, globe and anchor. These kids were proud and the looks on their faces couldn't be faked - the students stayed awake.

The students kept this recruiter busy talking about what it meant to be a Marine - not about the college fund. The last time I checked the Marine Corps was the only service that met its accession goals - maybe we need to think about what it is we're trying to sell…"

CHESTY'S TRUE SONS

By Smith Hempstone

"The road to hell is paved with the bleached bones of leaders who forgot to put out local security." – Chesty Puller

Any Marine reading this should already know all about Chesty Puller. For those who aren't familiar with the name, Chesty was what every leader should strive to be, and was so well loved by his Marines that even his jeep driver, a Sergeant named Orville Jones, became something of a celebrity simply because he had known the general. Puller had a son of his own, Lewis Jr., whose own story is contained in the Pulitzer Prize winning book "Fortunate Son" - but it can be argued that the strongest of bonds were those forged between Chesty and his Marines in the heat of combat on the battlefields of Guadalcanal, Peleliu and the Chosin Reservoir. One young officer in Korea said of Chesty, "I'd follow him to hell - and it looks like I'm going to have to."

At the conclusion of the Puller's aptly titled biography "Marine!," author Burke Davis relates a conversation which took place between Chesty and his wife a few years after the general had retired. She asked him if there was anything he would wish for, given that his career was over. His reply was, "I'd like to do it all over again. The whole thing. And more than that - more than anything - I'd like to see once again the face of every Marine I've ever served with."

They buried Chesty Puller at high noon on an apple-sweet October '71 day, the notes of taps thin and sad on the crisp

Tidewater air. The most decorated Marine in the history of the Corps did not hit his last beachhead alone.

They were all there: Chapman, Walt, Shoup, Greene, Silverthorne, Thomas, more than two dozen generals from a service in which stars come neither quickly nor easily. Pink-cheeked recruits down from Quantico and turkey-necked old timers, crackers who could remember how Chesty won his first Navy Cross against Sandino in Nicaragua forty years ago. More than fifteen hundred Marines and ex-Marines found their way to that remote churchyard in Virginia to pay final homage to a superlative fighting-man who in his own lifetime had become a myth.

The wonder of it all is that Lewis Burwell Puller lived either to make general, or to die in bed at the age of seventy-three. Haiti, Nicaragua, Guadalcanal, Peleliu, Inchon, Yongdongpo, Chosin Reservoir. At any of half a dozen places Puller might have left his bones to whiten with those of so many of the brave men he led. For Chesty led from out front and insisted that his officers do so, which is why his 1st Marines lost seventy-four percent of its officers and "only" sixty percent of its enlisted men in the caves of Peleliu's Bloody Nose Ridge.

But although Puller bore to his grave the scars of a dozen wounds, the God in whom he reposed such quiet trust denied him the battlefield death for which, in reality, he was born. Making general was another matter.

For in the service as in civilian life there's a small hello for a man with a salty tongue who is unafraid to use it on his superiors. Chesty always maintained that the "fat-assed generals" had it in for him and indeed, he did not win his first star until he had served thirty-three years, won an unprecedented five Navy Crosses and led his 1st Marines out of the "Frozen Chosin" carrying their dead and wounded, trailed by the shattered remnants of less-favored regiments, and better equipped (from material abandoned by other units) than when the Korean front collapsed.

Those "fat-assed generals" - or perhaps the Women's Christian Temperance Union (he always insisted Marines fought better on whiskey than on ice cream) - saw to it that the barrel-chested Puller never commanded anything larger than a regiment in combat. His third star was a "tombstone" promotion made on the occasion of his retirement because of his fifty-six decorations for valor, and they turned Puller down in 1965 when he tried to get recalled to active duty so he could go to Vietnam. Instead, his son went, and young Puller lost both legs and parts of six fingers in a landmine explosion.

To serve under Chesty was to have a good chance to die. And yet enlisted Marines, who are not given to the adulation of their generals, fought for the chance to follow him and came down out of the mountain hamlets of half a dozen states to bury him last week.

Curious. Or is it? It was much the same with that strange, harsh, God-fearing man, "Stonewall" Jackson, Puller's fellow Virginian. Jackson would have a hungry Confederate soldier shot for stealing apples in Maryland, and yet his butternut legions cheered him to a man whenever he showed himself. As Lee remarked sadly after Chancellorsville, Jackson's presence on the battlefield was worth that of two crack regiments. So it was with Puller.

No soldier ever loved the brilliant Douglas MacArthur, which leads one to the conclusion that enlisted Marines, with that curious intuition of unschooled men, realized two things: Chesty Puller was tougher than any of them, and despite and almost because of that he genuinely cared about them. He might - and almost certainly would - lead them into hell, but would be with them all the way and lead them out the other side, savoring their victories, and mourning their deaths, for they were all no less than his own blood and bone, Chesty's true sons.

The Services are experiencing a difficult period. Men not fit to shine the boots of Chesty Puller make a mockery of everything for which he stood. You have to go into the rural areas to find a post office outside which recruiting posters can stand undefaced. A man who has served his country in Vietnam, laying his life on the line, has to apologize to hairy stay-at-homes for his deeds.

And yet this too will pass. Ever since the world began there have been meat-eaters and grass-eaters, those who would fight and those who would rather talk. Both in the cities and in the boondocks, in the concrete hell of Spanish Harlem and in the grim coal mines of Harlan County, Kentucky, Marine recruiters are still finding raw-boned youngsters willing and eager to go through the valley of the shadow for a lantern-jawed, profane, compassionate man like Chesty Puller.

In the end, it matters little whether the rest of us understand or appreciate warriors such as Puller; there is little we can do to add or detract from what they have been and are. This nation could not exist without them, and all of us comfortably home today owe each of them an immense debt of gratitude. When the smoke of the last volley cleared over the grave of Chesty Puller, it would have been a small man who would not have conceded that.

Newspaper columnist and former Ambassador Smith Hempstone served with the 1st Marine Division during the Korean War and published this article on October 20, 1971.

LONG AND WINDING ROAD

"Being defeated is often a temporary condition. Giving up is what makes it permanent." - Marilyn vos Savant

For me running has always been a big part of being a Marine. I know there are many ways to stay in shape, but as far as I am concerned low impact exercises and those wacky devices you see on late-night TV infomercials are strictly for housewives, Air Force generals and Poppin' Fresh the Pillsbury Doughboy. I firmly believe it is important for a Marine to maintain his body in top physical condition at all times. After all, you never know when you might have to take a PFT, and believe it or not I still think that way despite being retired for a number of years.

Unfortunately for me, I was never was a particularly gifted runner. I worked hard at it and regularly achieved a respectable time, but no matter how much I trained I just couldn't keep pace with those naturally gifted individuals who could run like the wind. And it was between stints on active duty that I came to truly understand the importance of staying ready.

I will never forget my first day back on active duty at 1st Recon Battalion at Camp Pendleton. I had just driven cross-country from New York, and had spent a considerable amount of time on leave visiting my sister in San Diego. Needless to say, I wasn't getting much in the way of exercise. My first PT formation after reporting in concluded with the First Sergeant commanding myself and the rest of the platoon sergeants to take charge of our troops and "run 'em up Ball Buster!" I had no clue what that was, so the Marines in my platoon pointed at a steep hill looming behind us in the distance. We took off at a double time, and when we reached Ball Buster turned left and went straight up the front face. It turned out to be even higher and steeper than it had looked from a distance, and in fact there were some places where the slope was almost vertical. In my poor physical state it wasn't long before I was sucking wind, and I swore my heart was pounding so hard it was going to come through the front of my chest. Somehow I managed to survive, and that day I resolved to never again let myself get that far out of shape.

That resolution held true even while I was stationed in the Congo. It was tough to get in a run there, since on pretty much every route it was necessary to navigate over and around tons of trash and the bodies of the dead and dying. Even the smell was overwhelming at times. But we had to "suck it up" since PFTs were administered every six months just like back in the fleet. The one exception to that was the detachment in Mexico City, since the air quality there was deemed to be so bad that running was considered an unhealthy activity.

One of the events the Marine Corps uses to simultaneously promote physical fitness and the Corps itself is the Marine Corps Marathon. It is run every year in late October or early November, and has become one of the premier road races in the country. The route begins at Arlington Cemetery, goes the length of the Capitol Mall, and finishes in front of the Marine Corps War Memorial. It is a great event in every respect, although I do have one small criticism. For some reason they decided to call it "the People's Race" as part of some early promotion, and that just doesn't sound right to me. In fact, it sounds like it was written my Karl Marx.

It took until 1981 before I talked myself into running my first marathon, and of course it couldn't be just any marathon

- it had to be "ours." I signed up early along with some of my buddies and began to train hard - but unfortunately I sustained a serious knee injury about three months before the race, and subsequently began to make noises like a Chicago Cubs fan (i.e. wait until next year). When the time came for the marathon I made the trip down to D.C. with my buddies anyway, and since I had long since paid the entry fee I went ahead picked up my race packet and t-shirt.

Here's where the story really begins. The following morning I became so motivated watching the other Marines in my group prepare to run that I decided to participate in the start and "run for a little while until I got tired." After all, what could it hurt?

The start itself got me even more motivated. Instead of a starter's pistol they fired a blank round from a 105mm howitzer, and as that big cloud of smoke drifted across the Potomac I could feel the adrenalin start to empty into my stomach. As I started to shuffle along I began to converse with the girl running next to me, and I soon discovered she was a student at Georgetown and a huge fan of the Marine Corps. I can still hear her today, telling me how she admired Marines because they never gave up. It just added to the earlier motivation, so I decided against telling her about my injury because it sounded too much like an excuse - even though it was true. Before I realized what had happened we had reached the Capitol Building, which is roughly the halfway point. I said to myself "Wow, I made it this far. It's all downhill from here!"

At mile sixteen I really began to feel the effects of the long layoff. My legs felt like they were on fire, and every step was excruciating. It was then that I began to think about what I call the "Q" word. Quitting is something Marines just don't do, but I was beginning to wonder if I was going to be able to make it the last ten miles. At that point I decided I had to at least make it to the next aid station.

When I reached the water stop and first aid station at mile eighteen I was just about done. If I had been a character on *Seinfeld* I would have said, "Stick a fork in me, Jerry!" But then something happened which proved to me once and for all that it really is all mind over matter - if you don't mind, it doesn't matter. My buddies who had been working at the water stop saw me, and before I could collapse into the heap of human refuse I felt like they formed up behind me in their boots and cammies and began chanting my name. Talk about motivation! When that happened there was suddenly no doubt in my mind I would finish.

I left my friends behind in Arlington Park, and as I crossed the bridge over the Potomac I was going on willpower alone. There was nobody there to cheer me on, so I had to reach deep down inside to find my own motivation. I think the thing that kept me going was the same as that which kept Marines at Belleau Wood and Tarawa moving forward through enemy fire - pride in the Corps. I was wearing a t-shirt that said UNITED STATES MARINE CORPS on it, and I was damned if I was going to dishonor it.

As I came around that final corner I had nothing left. The muscles in my legs were swimming in lactic acid, and I was falling forward more than running. And then a strange thing happened. I looked up ahead and saw the Marine Corps War Memorial, that magnificent bronze statue of Marines raising a flag on Iwo Jima. I knew what I was going through in that marathon paled in comparison to what those Marines had endured, and suddenly my legs began to churn and I felt as if I was flying. I sprinted the last few hundred yards and crossed the finish at a dead run.

What happened that day had a profound effect upon me. The certificate I received for completing the marathon in five hours and three minutes had my name on it, but it should also have contained the names of Ira Hayes, Rene Gagnon, John Bradley and the rest of those who were immortalized in that statue. Later in my career I took to giving out dictionaries to some of my Marines (because I am a stickler for proper spelling and grammar) but before I did so I blacked out words like “quit” and “fail,” because such words have no place in the vocabulary of a United States Marine. After all, winners never quit, and quitters never win!

THE RIGHT STUFF

"The Marine pilots were superb. They would fly down a gun barrel." – LtCmdr Edgar Hoaglund USN, Phillipines 1945

Some people still don't understand why military personnel do what they do for a living. The famous exchange between Senators John Glenn and Howard Metzenbaum is worth reading again, because it perfectly illustrates the stark contrast between the selfish and the selfless in this nation. Not only is it a pretty impressive impromptu speech, it's also a good explanation of why men and women in the Armed Services do what they do for a living. But first consider Senator Glenn's résumé:

John H. Glenn Jr., a true American hero, grew up in the small religious town of New Concord, Ohio. The son of a World War One veteran, his childhood recalls a Norman Rockwell painting: Decoration Day parades, little kids playing in fields and woods, hot fudge sundaes at the local dairy, and marrying the girl next door.

He enrolled in the Naval Aviation Cadet Program in 1942 and served with the Marines' VMO-155 during World War Two, flying fifty-nine combat missions in F4U Corsairs over the Marshalls, and earning two Distinguished Flying Crosses. Glenn and the Marine fliers of VMO-155 arrived at Majuro in the Marshalls in July of 1944, after the heavy fighting in that area had subsided and the Americans had captured the large strategic atolls of Majuro, Kwajalein, Roi-Namur, Eniwetok, and Namu. But isolated Japanese forces still held out on Wotje, Maloelap, Mili, and Jaluit.

VMO-155's job was to keep the Japanese forces suppressed, to prevent them from staging any counter-attacks by air or water. Glenn's first combat mission took place a few days after he landed; it was flak suppression. Fly some Corsairs over Maloelap and blast away at any anti-aircraft installations that opened up. Not exactly glamorous, but very real. On this first mission Monty Goodman, a wise-cracking flier from central Pennsylvania and one of Glenn's good friends, didn't make it back to the rendezvous point.

In November Glenn's squadron moved over to Kwajalein, where they continued to attack the Japanese forces in the Marshalls. Now they had a new weapon, napalm, which would only become infamous twenty-five years later in Vietnam. It was a hideous weapon, and they used it "where intelligence thought there were a lot of people. It was terrible to think what it was like on the ground in the middle of those flames… it made you think. Then the psychology of war took over. We were fighting in a war we hadn't started, for the survival of our country, our families, our heritage of freedom."

Glenn left the Marshalls in early 1945, and returned stateside. For the last few months of the war he was at Pax River, test flying planes like the F8F Bearcat and the Ryan Fireball FR-1. Promoted to Captain by war's end, he decided to make a career of the Marines.

After checking out in the F9F Panthers that the Marines were using in Korea Glenn flew there in February 1953, and was assigned to the First Marine Air Wing's VMF-311, based at P'ohang. Two things immediately struck Glenn about Korea: the cold and "kimchi," a Korean staple consisting of fermented cabbage, onion, radishes, and garlic. It actually solidified during it fermentation, and "if you were downwind when someone had the kimchi jug open, the smell wasn't something you'd forget." (Did you ever hear the expression, "You'll be in deep kimchi!"? That's the stuff.)

P'ohang was about 180 miles from the front. Armed with three thousand pounds of bombs and five-inch HVARS (High Velocity Aircraft Rockets), the heavily-built F9F Panthers were well-suited for ground attack missions. They flew constantly, providing close support for the Marines at the front. Glenn's good friend Tom Miller and other experienced pilots had advised him to steer clear of "flak traps," and they had orders against making a second run at a target - but like all of us, Glenn sometimes had to learn the hard way. One day flying over Sinanju he spotted a North Korean anti-aircraft gun emplacement, noted its position, and circled back to blast away at it with the F9F's four 20-mm cannon. His Panther got hit in the process, and he could hardly keep the plane flying, constantly pulling back on the stick just to keep it level. He made it back to P'ohang to find a "hole in the Panther's tail that was big enough to put my head and shoulders through. There were hundreds of smaller shrapnel holes around the big one. We figured it was a thirty-seven millimeter shell that hit me; a larger one would have blown the tail off. Crews replaced the tail and the Panther flew as good as new. That was the last time I went in for a second run." A week later he got hit again; this time an even larger anti-aircraft shell had blown the napalm tank off his wing, and while he landed safely, that plane was toast.

Glenn summarized this part of his Korean War experiences in *John Glenn: A Memoir*: "I enjoyed the kind of air-to-ground combat we were doing. Flying in support of ground troops is what had attracted me to the Marines when I heard about Guadalcanal way back at Corpus Christi. Marines look at themselves as a team... But I also hoped for air-to-air combat. That was the ultimate in fighter flying, testing yourself against another pilot in the air. Ever since the days of the Lafayette Escadrille during World War I, pilots have viewed air-to-air combat as the ultimate test not only of their machines but of their own personal determination and flying skills. I was no exception. You believe you're the best in the air. If you do, you're not cocky, you're combat-ready. If you don't, you'd better find another line of work."

After flying sixty-three missions in a Marine Corps F9F Panther from airbase K-3 at Pohong Dong (P'ohang), Glenn applied to fly F-86 interceptors with the Air Force on an exchange program. He was assigned to the 25th FIS (Fighter Interceptor Squadron) at K-13, Suwon, where the 51st Fighter Interceptor Wing was headquartered. They patrolled the area just south of the Yalu, the so-called "MiG Alley," in long figure eights, always turning towards the north to keep from being surprised. The F-86 Sabres and the MiGs were evenly matched. Both had 6,000-pound thrust jet engines, and could go supersonic in a dive. The MiG was smaller, and it could climb higher and faster. The Sabre was faster in level flight and in a dive, had a greater range, and could turn tighter in a fast dive. The Sabre carried six 50-caliber machine guns, while the MiG relied on a single 37-mm and two 23-mm cannon.

Unlike the Marines, the Air Force pilots tended to fly the same plane day after day. It became "their" plane, and nose art and other personal decorations flourished. Not long after Glenn began flying his F-86F Sabre the fuselage sported in large script: *LYN ANNIE DAVE*, for his wife and two kids. After enough of his moaning about the absence of MiGs, he went out to the flight line one morning to find a big red M painted on it, with letters trailing off it, so it read: "Mig Mad Marine."

Soon the USAF Sabres were ordered to fly ground attack missions if they were returning from unsuccessful MiG-hunting with a full load of munitions. On such a raid over Sinanju Glenn's CO was lost, and as a result Glenn began leading two and four-plane flights. Now he would be "the shooter." On July 12, 1953 he was flying with 1stLt Sam Young on his wing when he spotted a MiG and chased it forty miles into Manchuria. The rules of engagement permitted UN fliers to cross the Yalu when "in hot pursuit." Abruptly the MiG slowed to land, and Glenn opened up with his six .50s. The bullets lit up the fuselage and wing, sending up bright sparks. Flames burst out, and as the MiG hit the ground it exploded. Glenn flew low enough to see the MiG spread out over a hundred yards. He rendezvoused with Young, and flew back to K-3 for an impromptu celebration.

A few days later he got the chance to mix it up with some more MiGs when his flight of four F-86s was bounced by sixteen MiGs. Soon four other Sabres joined the fray, and a WWI-style dogfight ensued - only the planes were flying at 600 MPH instead of 100 MPH! Glenn's wingman on this day, Jerry Parker, scored some hits, but was soon hit himself. He broke off to escort Parker back to K-13. Six MiGs came after them, and Glenn's only choice was to "light up the nose," fire at them from long range, in the hope they would break off their attack. They did, and then Glenn went after them in earnest, catching up to the tail-ender and flaming it. "The MiGs' tactics were so poor I could only imagine it was a training flight, or they were low on fuel, but we were unbelievably lucky."

Three days later he downed his third MiG. It would be his last of the war. There were a few more days of bad weather, and then the armistice was declared. He had flown twenty-seven Sabre missions with the USAF 51st FIW, and earned another DFC and eight Air Medals in Korea.

After the Korean War Glenn entered the Navy's prestigious Patuxent River Test Pilot School (universally known in the military flying community as "Pax River"). He rose to the rank of Major in the Marine Corps after three years in test flight, and in 1957 became a minor celebrity when he flew the first supersonic, trans-continental flight - a project that he devised and managed himself. Flying a Vought F8U Crusader, he developed the plan to fly from Los Angeles to New York at an average speed above Mach 1, which required three aerial refuelings from flying tankers. He completed the flight in three hours and twenty-three minutes.

In 1958 John Glenn was selected as one of the original seven Mercury astronauts. As portrayed in Tom Wolfe's "The Right Stuff," he was the clean-cut, go-getter of the group. While he was not chosen to fly either of the first two flights, as it worked out his flight, the third in the Mercury program, was the choice mission. It carried him far beyond its three orbits and five hours to global fame: a Broadway ticker tape parade, a meeting with President Kennedy, and an eventual career in politics as U.S. Senator for Ohio. I loved reading about his famous speech – "Yes, I've held a job, Howard" - when his opponent, Howard Metzenbaum, a self-made millionaire, accused Glenn, as a lifetime "government employee" of never having held a real job. This IS a typical, though sad, example of what some who have never served think of the military. In an inspired turn, an incensed Glenn turned it around, and used his response to trumpet his military service and to proclaim the dignity, honor, and sacrifice of military service:

Senator Metzenbaum to Senator Glenn: "How can you run for the Senate when you've never held a 'real' job?"

Senator Glenn: "I served twenty-three years in the United States Marine Corps. I served through two wars. I flew 149 missions. My plane was hit by antiaircraft fire on twelve different occasions. I was in the Space Program. It wasn't my checkbook, Howard; it was my life on the line. It was not a nine to five job, where I took time off to take the daily cash receipts to the bank.

I ask you to go with me... as I went the other day... to a Veterans Hospital and look those men - with their mangled bodies - in the eye, and tell THEM they didn't hold a job!

You go with me to the Space Program at NASA and go, as I have gone, to the widows and orphans of Ed White, Gus Grissom and Roger Chaffee... and you look those kids in the eye and tell them that their Dads didn't hold a job.

You go with me on Memorial Day and you stand in Arlington National Cemetery, where I have more friends buried than I'd like to remember, and you watch those waving flags. You stand there, and you think about this Nation, and you tell ME that those people didn't have a job?

I'll tell you, Howard Metzenbaum, you should be on your knees every day of your life thanking God that there were some men - SOME MEN - who held a REAL job. And they required a dedication to a purpose - and a love of country and a dedication to duty - that was more important than life itself. And their self-sacrifice is what made this country possible. I HAVE held a job, Howard! What about you?"

ESPRIT DE CORPS

Author Unknown

"There is no better group of fighting men anywhere in the world than in the Marine Corps." – Senator Irving M. Ives

Ask a Marine what's so special about the Marines and the answer would be "esprit de corps," an unhelpful French phrase that means exactly what it looks like - the spirit of the Corps. But what is that spirit, and where does it come from?

The Marine Corps is the only branch of the U.S. armed forces that recruits people specifically to fight. The Army emphasizes personal development (an army of one), the Navy promises fun (let the journey begin), and the Air Force offers security (it's a great way of life). Missing from all of these advertisements is the hard fact that it is a soldier's lot to suffer and perhaps to die for his people, and to take lives at the risk of his own. Even the thematic music of the services reflects this evasion.

The Army's *Caisson Song* describes a pleasant country outing over hill and dale, lacking only a picnic basket. *Anchors Aweigh*, the Navy's celebration of the joys of sailing, could have been penned by Jimmy Buffet. The Air Force song is a lyric poem of blue skies and engine thrust. All is joyful and invigorating, and safe. There are no landmines in the dales nor snipers behind the hills, no submarines or cruise missiles threaten the ocean jaunt, no bandits are lurking in the wild blue yonder.

The Marines' Hymn, by contrast, is all combat. We fight our country's battles, first to fight for right and freedom, we have fought in every clime and place where we could take a gun, in many a strife we've fought for life. The choice is made clear. You may join the Army to go to adventure training, or join the Navy to go to Bangkok, or join the Air Force to go to computer school. You join the Marines to go to war. But the mere act of signing the enlistment contract confers no status in the Corps. The Army recruit is told from his first minute in uniform that "you're in the Army now, soldier." Navy and Air Force enlistees are sailors or airmen as soon as they get off the bus at the training center. The new arrival at Marine Corps boot camp is called recruit, or private, or worse (much worse), but not Marine. Not yet; maybe not ever. He or she must earn the right to claim the title, and failure returns you to civilian life without hesitation or ceremony.

My recruit platoon, Platoon 2210 at San Diego, California, trained from October through December of 1968. In Vietnam the Marines were taking two hundred casualties a week, and the major rainy season operation, Meade River, had not even begun. Yet our drill instructors had no qualms about winnowing out almost a quarter of their 112 recruits, graduating eighty-one. Note that this was post-enlistment attrition; every one of those who were dropped had been passed by the recruiters as fit for service. But they failed the test of boot camp, not necessarily for physical reasons (at least two were outstanding high-school athletes for whom the calisthenics and running were child's play). The cause of their failure was not in the biceps or the legs, but in the spirit. They had lacked the will to endure the mental and emotional strain, so they would not be Marines. Heavy commitments and high casualties notwithstanding, the Corps reserves the right to pick and choose.

But the war had touched boot camp in one way. The normal twelve-week course of training was shortened to eight weeks. Deprived of a third of their training time, our drill instructors hurried over, or dropped completely, those classes without direct relevance to Vietnam. Chemical warfare training was abandoned. Swimming classes shrank to a single familiarization session. Even hand-to-hand combat was skimped. Three things only remained inviolate: close order drill, which is the ultimate discipline builder; marksmanship training, the heart of combat effectiveness; and classes on the history, customs and traditions of the Corps.

History classes in boot camp? Stop a soldier on the street and ask him to name a battle of World War One. Pick a sailor at random to describe the epic fight of the Bon Homme Richard. Everyone has heard of McGuire Air Force Base, so ask any airman who Major Thomas B. McGuire was, and why he is so commemorated. I am not carping, and there is no sneer in this criticism. All of the services have glorious traditions, but no one teaches the young soldier, sailor, or airman what his uniform means and why he should be proud to wear it. But ask a Marine about World War One, and you will hear of the wheat field at Belleau Wood and the courage of the Fourth Marine Brigade. Faced with an enemy of superior numbers entrenched in tangled forest undergrowth, the Marines received an order to attack that even the charitable cannot call ill advised. It was insane. Artillery support was absent and air support hadn't been invented yet, so the Brigade charged German machine guns with only bayonets, grenades and indomitable fighting spirit.

A bandy-legged little barrel of a Gunnery Sergeant, Daniel J. Daly, rallied his company with a shout. "Come on, you sons of bitches! Do you want to live forever?" He took out three of those machine guns himself, and they would have given him the Medal of Honor except for a technicality. He already had two of them. French liaison officers, hardened though they were by four years of trench bound slaughter, were shocked as the Marines charged across the open wheat field under a blazing sun and directly into enemy fire. Their action was so anachronistic on a twentieth-century battlefield that they might as well have been swinging cutlasses. But the enemy was only human; they couldn't stand up to this. So the Marines took Belleau Wood. Every Marine knows this story, and dozens more. Every Marine is taught them in boot camp as a regular part of the curriculum.

You can learn to don a gas mask anytime, even on the plane en route to the war zone, but before you can wear the emblem and claim the title you must know of the Marines who made that emblem and title meaningful. So long as you can march and shoot and revere the legacy of the Corps, you can take your place in the line. And that line is unified in spirit as in purpose. A soldier wears branch of service insignia on his collar, and metal shoulder pins and cloth sleeve patches to identify his unit. Sailors wear a rating badge that identifies what they do for the Navy. Marines wear only the eagle, globe and anchor, together with personal ribbons and their cherished marksmanship badges. There is nothing on a Marine's uniform to indicate what he or she does, nor (except for the 5th and 6th Regiments who wear a French Fourragere for Belleau Wood) what unit the Marine belongs to. You cannot tell by looking at a Marine whether you are seeing a truck driver, a computer programmer, or a machine gunner. The Corps explains this as a security measure to conceal the identity and location of units, but the Marines' penchant for publicity makes that the least likely of explanations. No, the Marine is amorphous, even anonymous (we finally agreed to wear nametags only in 1992), by conscious design.

Every Marine is a rifleman first and foremost, a Marine first, last and always. You may serve a four-year enlistment or even a twenty-year career without seeing action, but if the word is given you'll charge across that wheat field. Whether a Marine has been schooled in automated supply, or automotive mechanics, or aviation electronics, is immaterial. Those things are secondary - the Corps does them because it must. The modern battle requires the technical appliances, and since the enemy has them, so do we. But no Marine boasts mastery of them. Our pride is in our marksmanship, our discipline, and our membership in a fraternity of courage and sacrifice. "For the honor of the fallen, for the glory of the dead," Edgar Guest wrote of Belleau Wood, "the living line of courage kept the faith and moved ahead."

They are all gone now, those Marines who made a French farmer's little wheat field into one of the most enduring of Marine Corps legends. Many of them did not survive the day, and eight long decades have claimed the rest. But their action has made them immortal. The Corps remembers them and honors what they did, and so they live forever. Dan Daly's shouted challenge takes on its true meaning - if you hide in the trenches you may survive for now, but someday you will die and no one will care. If you charge the guns you may die in the next two minutes, but you will be one of the immortals. All Marines die, in the red flash of battle or the white cold of the nursing home. In the vigor of youth or the infirmity of age all will eventually die, but the Marine Corps lives on. Every Marine who ever lived is living still, in the Marines who claim the title today. It is that sense of belonging to something that will outlive your own mortality that gives people a light to live by and a flame to mark their passing. Marines call it Esprit de Corps!

CHAPLAIN'S THOUGHT

By LtCol Johnson, Chaplain, USAF

"Light is the task where many share the toil." – Homer

Last Thursday morning I was one of more than three hundred runners in the NSA Armed Forces Week 5K run at Fort Meade, MD. It was pretty crowded at the start, but things thinned out after about five minutes or so, and I took my bearings. Perhaps two hundred yards ahead of me was a group of maybe eight or so Marines who were obviously running together. I decided that a good goal would be to beat them, which seemed reasonable as I am a macho Air Force Chaplain and they were only a bunch of United States Marines. I kept them in sight for the next couple of miles, but the longer the race went on, the younger those guys got. It became apparent to me in the last half mile that I was not going to catch them, and I resigned myself to finishing well behind.

Then I noticed that one of their number was struggling and was gradually dropping off the pace. I panted out a word of encouragement as I caught him and realized that even though he was injured he was not about to give up. Within one hundred yards of the finish line I then saw a strange sight. The entire group of Marines had made a u-turn in the road and were running back towards me. As they ran past me I noted their well-chiseled muscles and the determined set of their jaws. I glanced over my shoulder in time to see them rally around their buddy to provide the emotional support of the team so that they could all finish together.

I was impressed. No way would they leave a struggling comrade behind. As I entered the finishing chute I murmured a prayer. "God, I'm glad those guys are on our side." And so it was that I learned a theological truth from the U.S. Marines that is as vivid as any my seminary professors ever taught. "If anyone... sees his brother in need but has no pity on him, how can the love of God be in him? Let us not love with words or tongue but with actions and in truth."

Last Thursday I witnessed "a few good men" in action. They reminded me of the strength of being a team, and that words without actions are pretty much useless.

Thanks, Marines!

SO THERE!

By Drew Carey

"Nobody ever promised you a rose garden."
– USMC Recruiting Poster

A lot of people don't realize it, but comedian Drew Carey was once a Marine. He tends to downplay his time in the Marine Corps Reserve, and while he certainly is an unlikely specimen I have no doubt deep down inside he still feels the pride of being one of the world's finest. As proof I offer his take on political correctness, sexual harassment and the Tailhook incident:

How many militant feminists does it take to change a light bulb? Two. One to change the bulb, and one to kiss my ass. That's right. I said kiss my ass. 'Cause I've had it. I'm tired of being pushed around. Tired of being grouped in with all the deadbeat dads and rapists and lecherous bosses just because I'm a man. All men aren't "potential rapists." I'm not a potential rapist.

But, I am a potential murderer if all of you don't shut up and get out of my face already. You've ruined it for everybody. Everybody, do you hear me? Men, women - everybody. Because of you and everyone else in this society that needs to play political victim and go to court instead of just dealing with it themselves, no one can have any kind of fun anymore. Men and women can't flirt, or hug, or look at anyone sideways because of you and your lawyers.

Are you happy? You've used a stink bomb to kill a few ants. And while I'm at it... Naval Aviators, who are willing to die so that we can have low prices at the gas pump, should be able to throw the wildest parties they can manage without one uptight biddy coming in and stopping it.

There were scads of women at that Tailhook party who were having the time of their lives, voluntarily being just as debauched as any of the men were. Everyone who flew a plane, or even knew someone who flew a plane, knew how wild those parties were and what went on. What did she expect? A prayer service?

And why didn't she just throw some punches of her own when these couple of guys groped her? Why didn't she give them what they had coming and just kick them in the balls? Didn't our tax money go to teach her how to fight?

I'm not trying to make the idiotic "she had it coming" argument here, which would go something like "of course they grabbed her breasts, look how big they are." Plus, just reaching out and grabbing some boob is wrong no matter what. When I was in college, even at our most drunken fraternity parties we never acted like that. No matter how hard I try I can't think of an excuse good enough to do something like that. But it's still nothing to lose a career over.

Besides, fighter pilots are supposed to be aggressive assholes. That's what we pay them for. I don't know about you, but I don't want a Navy full of fighter pilots who are gifted at giving sensitivity seminars. I want mad-dog, rabid killers going to battle for me and mine. Man or woman.

When our stable gas prices are threatened by a Middle-Eastern Madman, when we want to force our form of government on some poor, unsuspecting Latin American country, when uppity foreign diplomats "forget" to pay their parking tickets, I want to be able to call on men and women who like to fight and drink.

I want a Naval officer who knows how to whack some drunk in the balls when he grabs her tits, not call a press conference, and a lawyer. If you're a wimp who doesn't know how to find the exit at a rowdy party, go fly a kite, not a jet fighter. So there!

MARINE TRAINING

"Discipline yourself and others won't need to."
- John Wooden

As I pointed out in Swift, Silent and Surrounded,' the habits we acquire in the Marine Corps tend to stay with us for our entire lifetime. Sailors tend to lose their nautical perspective within a year or two, soldiers are assimilated into the civilian mainstream in a matter of weeks, and the Air Force - well, they tend to revert to their pre-enlistment ways seconds after being handed their discharge papers. Why is that? Some would say we Marines are brainwashed, and have had the lessons of the Corps pounded deep into our psyche through constant repetition and high stress. That is, of course, part of it. But the real reason, I believe, is that something hard won is not easily relinquished. We have each earned that Eagle, Globe and Anchor, and our subconscious will never let us forget it. Here is one Marine's perspective:

I've been out of the Corps a year longer than I was in (in four, out five), and I still don't carry anything in my right hand, unless it's absolutely necessary. After all, you never know when you'll have to salute someone.

"The Marines' Hymn" still gives me cold chills, and a picture of Mount Suribachi brings a tear to my eye.

I always stand at attention for the national anthem, with hand over my heart. I don't put my hands in my pockets when walking, and walking in step is a must.

A rack is still a rack (not for hanging hats), a head is still a head (not the one on your shoulders), and the deck is still the deck (we're not talking sailboats, either).

At the office co-workers think I'm crazy, using terms like guard mail (instead of interoffice correspondence), direct order (instead of directive), and locked on (instead of understood). The task at hand is always a "mission," and no mission is ever too tough.

Even the days aren't long enough. Not that I complain about a 9-to-5 job, or working regular hours; I don't. But it seems that others - civilians - are always complaining about how hard and/or horrible their work is. Get real. Join the Marine Corps....

A headache, stomach ache, or cold might keep the average employee home. Calling in sick, except in case of rare disease or disaster, is out of the question for a Marine. Being late is equally unsat. (What's that? Ask a Marine.)

The word "Sir" involuntarily rolls off my lip when addressing senior management. Some think it's great; others don't care for it at all. (Remember the first sergeant's cry? "Don't call me 'Sir', I work for a living!") At any rate, I find myself explaining that its "ingrained Marine Corps training," which is always a door opener for further conversation.

Such a statement can also be beneficial during other interactions, such as those with police officers. Fortunately, my experience in that area is limited, but any mention of "Marine" is usually a good icebreaker and lead in to conversation about the Corps. It seems that there's a mutual respect between the Police and the Marines; in fact many are Marines (former and reserve). Not everyone can be a Marine, and if you are, say so. A Marine bumper sticker in the window and dog tags hanging in the rearview mirror can also go a long way.

Speaking of bumper stickers, have you ever noticed how many there are out there? Hundreds of thousands, perhaps millions, proudly displayed on cars and trucks from New York to California. Marines are everywhere.

And when they're not in their bumper-stickered vehicles, you can otherwise spot them in their bright, red and gold USMC jackets, caps (not hats), T-shirts, and other assorted accessories. But not all Marines are that easily recognizable. Some garb is understated in black, silver, green, or camouflage. Designs range from a simple Marine Corps emblem, to the Tasmanian Devil or a leatherneck tattoo, to an elaborate display of Marine weaponry. Sayings may include "Once a Marine, Always a Marine," "Semper Fi" (do or die), or any variation thereof. The words may be different, but the theme is always "Marine."

Marines will proudly inform you, and anyone else who happens to be listening, that they were in the Corps. Their comment may have no connection with the present conversation or situation, at least not to the common ear, but anything can, and will, rouse memories in a Marine. You could be in a crowded doorway, taking refuge from a storm, and a forty-something gentleman tells you he doesn't need an umbrella because he was a Marine, and compared to the monsoons in Southeast Asia or Okinawa, this downpour is just a sprinkle.

Or the moving man mentions in passing that he developed strength and endurance in the Corps. And there's the real estate agent who points out with pride his previous service when you pass by the local Marine monument.

From city to city, women in grocery lines and beauty parlors tell stories about their children, grandchildren, nephews and nieces who are, or were, Marines. From barroom to bowling alleys, from boat to backyard barbecue, fathers and grandfathers vividly recall life in the Corps to anyone who will listen.

Marines will seize any opportunity to volunteer information about their adventures in the Corps. They may casually note their branch of service, or unload an entire bag of sea stories. Fortunately most folks don't mind - unless, that is, they find themselves in the company of two or more Marines. In that case, they can forget getting in a word edgewise.

And remarkably, Marines always seem to find each other. In the midst of any crowd, two leathernecks will somehow get together, and when they do, it's an instant reunion. Forget formal introductions; these men are brothers. Call it "Marine bonding."

I recently attended a business conference (not Marine related) and found myself at a roundtable discussion. Actually, it was a luncheon, but the conversation was supposed to be business. Somehow, someone mentioned "Marine," and the gears immediately, and permanently changed. Another gentleman, who also happened to be a Marine, wanted to know what battalion, when, where served, with whom, how long. Of course, he too, was asked to share his case history.

None of the other people at the table, which included a Navy corpsman, Army sergeant major, and Air Force pilot, could compete. In fact they tried to offer tidbits about their service, but a mere "Oh really?" or "That's nice" was the only reaction they could get from the Marines. Interestingly, the non-Marines didn't seem to be perturbed. They were too busy listening to the sea stories.

Occurrences like these are not rare. In fact, they're probably the norm. Esprit de corps transcends the barriers of time and space, religion, and race. A Marine is a Marine. Once a Marine, Always a Marine. It's training you never outgrow, and a brotherhood you never forget.

THE CRUCIBLE

By Cameron McCurry

"Whether you believe you can do a thing or believe you can't - you are right!" – Henry Ford

It has been said that the "Old Corps" is the Marine Corps you belonged to five minutes before the guy you are talking to joined, and I guess that's true to some extent. Marines are both territorial, and traditional - we are set in our ways. Even so, as an organization the Corps is incredibly innovative. We are always coming up with new ideas across the spectrum. But as individuals, we tend to resist change. It was always better back in the "Old Corps." I am no exception.

When they decided to put name tapes on our camouflage uniforms I waited until the last possible day do it. I felt, as did most of my contemporaries, that not having to wear them was something which set us apart from the other three services. After all, I didn't need a piece of sewn on cloth to remind me of my own name. It was no different when the venerable M-14 rifle was phased out in favor of the lighter M-16. Marine marksmen longed for the heavier, larger-caliber, wooden-stocked weapon, saying it was more accurate and more dependable. Years later the same thing happened when the .45 caliber pistol gave way to the 9mm Beretta. How were we supposed to kill people with such itty bitty rounds? But the changes were made, and we got used to them... albeit kicking and screaming.

One of the things which sets the Corps apart from the other services is our approach to basic training. Marine

Corps boot camp is far more demanding than any of the others, and it always has been that way. So, how to improve upon perfection? Here is what the Corps came up with, and after the usual resistance we got used to this one too. As a member of the "Old Corps" I of course never experienced this new twist, so I will let my young friend Cameron McCurry explain:

The last thing you face as a Marine Recruit is the Crucible. General Krulak created it to give recruits a test of their endurance, leadership and ability to work as a team. The Crucible is tough. t begins on a Thursday morning at about 0230-0300. You get only a couple of hours of sleep each night, and your waking hours are a blur of motion and activity. You are given only two and a half meals to last you until Saturday morning when you face your final challenge. You do get a little hungry!

You face several obstacles on the Crucible. They are named after Marines who have won the Medal of Honor for their actions. As you go through the course, you learn about them and learn something new about Marine Corps history. The lessons are important, perhaps more important than the obstacles themselves. They serve as a reminder of why you put yourself through such training, and that someday you might be asked to make a sacrifice like they did.

I have to admit I have forgotten most of the obstacles that I went through, but have very clear memories of long forced marches, my feet being covered in blisters, and going through days with little sleep. There are two parts in particular that I can remember well.

The first thing was on Friday night. We were sitting down as a Company and told that we were going to wait at this one spot for a nighttime assault. With those hours, we managed to finish what was left in our MREs (the packaged meals that we had) and talk with each other as well as other platoons. This was the first time in training when we were allowed to just relax. It felt better than words can describe. I spent most of that time with my combat boots off trying to feel some sensation in my feet other than pain.

The assault began right after the sun went down. We started off in an Armored Personnel Carrier and waited for a signal. We were carrying boxes filled with sand to simulate ammo, and were tired beyond belief. When the signal came we sprinted like hell across an obstacle course that had barbed wire, walls, and fireworks going off at random intervals.

At one point a flare was set off, which was our cue to hit the deck. When I hit the ground, my earplugs came loose and I found myself a few short yards from where they were setting off the fireworks. I wasn't allowed to move, so I had to just lie there and listen to the explosions. Not pleasant to say the least.

The second most memorable part for me was on the last morning. We had maybe three hours of sleep at the most when we were dragged out of our tents and made to throw our packs on. This was it; we were finally on the last test that we had to face. We had to march several miles to a hill known affectionately as "The Reaper."

One thing is rather interesting about hiking early in the morning when you are sleep deprived. You start seeing things that aren't really there. I had the pleasure of seeing swirling colors in my field of vision. We were all rather relieved when the sun came up.

The Reaper was an event in itself. The hill is steep and it taunts recruits. You make it to the top sweating and sore, only to find that it drops down and goes back up again. This happened four or five times. But that last drop is where you see the flags of the States in their positions. At this point, you forget about your fatigue and run like hell because that's the goal.

We stood there in formation, most of us weeping in disbelief. The flag was raised and then the song "Proud to be an American" was played for us. At that point our Drill Instructors went up to each and every one of us. Sergeant Daniel stood in front of me with a smile on his face.

"We were worried about you for a while, but you hung in there. You kept at it even when things got bad, and that is what this is all about." He placed my Eagle Globe and Anchor into one hand and shook the other.

"Congratulations, Marine."

I think that no matter how many years pass, I am going to remember that moment very clearly.

Cameron McCurry is a Corporal in the Marine Corps Reserve

TRAINED TO DIE

By Colonel David H. Hackworth

"To win a war quickly takes long preparation."
- Latin proverb

Military staffers are busier than BX cashiers on payday, evaluating the lessons learned from the recent fireworks in Iraq. And that process is important. The stakes are too high not to get this long-term fight with terrorists dead right.

A case will soon be made for smart hardware and weapons to at least partially replace the current level of active-duty soldiers. While the right smart stuff is, of course, the way to go, if Cold War submarines and fighters such as the F-22 aren't culled from the weapons cache, it might well be goodbye ground troopers and the continuation of contracts for too many obsolete, gold-plated war toys. And it will be happy days for the war merchants, their always-available-at-the-right-price porker pals and the bean counters, who are more into systems than warriors.

Hopefully, our system of checks and balances will kick in, and Congress will ask if the wars in Afghanistan and Iraq are correct models to use for drastic changes to our force structure before our Secretary of Defense and his civilian slashers seriously weaken the time-tested force that defends our country.

Let's face it: Not only were both ragtag enemy armies incapable of really fighting back, vast money transfers also convinced many Afghani and Iraqi senior commanders to cut and run. And with Iraq, it wasn't just a thirty-day bombing campaign that prepared the field. So many missions - more than half a million - were flown over that country in the decade before we pre-empted Saddam Hussein from doing whatever he planned to do with his infamous inventory of yet-to-be-found doomsday weapons, it was a surprise to some that there were any targets left to "Shock and Awe."

But, as always after major operations, there are scores of basic lessons that must not be ignored, no matter what surgery the Pentagon bureaucrats and dilettante reformers ultimately perform. And one critically important lesson that has nagged at me for years is the inability of our Joes and Jills who bring up the rear to fight as infantry.

Since George Washington, all U.S. Army soldiers have always been trained first as riflemen. That skill has kept a lot of people in the rear with the gear alive and won a lot of fights, from our War of Independence to Korea - where then-Lieutenant Lloyd "Scooter" Burke led his unit's cooks in a counterattack that saved his company - to Vietnam - where then-Lieutenant Colonel Hank "The Gunfighter" Emerson dispatched his battalion's clerks and mechanics to save a company of besieged paratroopers.

In Iraq this time around, there were no neat front lines. The guerrilla enemy was everywhere - ambushing convoys and striking hard at our Army's soft underbelly. And many of these attacks proved the fallacy of one of the U.S. Army's frequently touted maxims: "We fight as we train."

Too large a number of Army rear-echelon folks failed the course when put to the test because they weren't trained to fight as grunts in Initial Training or when they joined their regular units. In many non-combat units today, this kind of live-or-die training gets brushed off by leaders who say, "Who needs this grunt stuff – we're ordnance, maintenance or transportation." Even during large training exercises, these vital survival skills are too often given only lip service.

No question the 507th Maintenance Company could've used the "more sweat on the training field, less blood on the battlefield" infantry training on that shameful day when its nine-vehicle convoy of ordnance troops took a wrong turn and bumped into a small enemy force in two pick-up trucks. Gun for gun, the 507th outnumbered the Fedayeen but still got clobbered to the tune of nine dead and five prisoners of war. Few 507th soldiers fired back because their weapons were clogged with dust. Hello? A soldier's weapon on a battlefield clogged with dust?! And those who weren't killed or captured straightaway ran liked scared jack rabbits - led, sadly, by their fleet-footed captain.

Congress is presently investigating this sorry display of cowardice and incompetence. Let's hope it has the smarts to conclude that the Army must return to the standard where every soldier truly is a rifleman first. The Marines still follow this rule, and when their support units in Iraq bumped into stay-behind fanatics they did what Marines have been doing well since 1775: killed the suckers and moved on!

RATS!

By Charles L. Fontenay

Guadalcanal, at the time I was there, was a tropical paradise of sticky heat, coconuts, mud, land crabs, Japanese bombers that drove us to the fox holes every night or so... and rats. Our contingent of military novices was housed in tents at Lunga Beach, and the rats considered our tents to be their nightly equivalent of Disneyland.

Some of those rats were as big as your average house cat, and they would climb the ropes and dance on the tent all night. They were also fond of its interior - the ground floor, literally - and we were careful to tuck up the blankets on our coats at night lest we find ourselves with unwanted bedfellows. The cot on which we kept our luggage also was kept free of any dangling participles that the rats might climb.

The rats were not afraid of us mere humans. One night when I was writing a letter on an upturned box I looked up to see one seated on its haunches, surveying me. I squashed it under one of my big GI shoes. In the tent with me were a closely knit trio: mustachioed McBane, Fuad Hanna, who was the image of a classical Egyptian and an inveterate gambler, and Lee Jones. They were jokers on occasion, and they'd pulled something on me (I forget what) that definitely called for revenge.

Fortune caused the PX to receive a supply of cheese niblets, and I stocked up on them. One evening, when the trio had gone to one of the island's outdoor movies, I prepared a proper reception for my first target, McBane, whose cot was across the tent from mine.

I dropped a couple of towels from the baggage cot to the ground - on its outside edge, where they were invisible, and ran a towel across the space between the baggage cot and McBane's cot. Then I sprinkled cheese niblets liberally on the baggage cot and on McBane's cot, under his blankets.

Shortly after our happy delegation returned from their movie, we all extinguished our lanterns and went to sleep. At least I did. I was awakened some time later by an unidentifiable noise, and for some intuitive reason, sitting up, I shone my flashlight across the tent.

McBane was sitting up on his cot, hugging his knees, an expression of utter dismay on his face.

"What's the matter, Mac?" I asked. "Can't you get to sleep?"

"It's the rats!" he informed me in a woebegone tone. "How in the hell they got up on my cot, I don't know, but they've been running all over me - all over my face. And their little feet are cold!"

Mark up one notch in the scheme of revenge. I went on back to sleep, and when I awoke the next morning, poor McBane was cramped up in the jeep outside the tent. He'd slept there all night to keep away from the rats.

Later, when we moved to floored and screened tents across the road from Henderson Field, I worked out a plan to get my next victim...

FELLOW MARINES

By Edward Andrusko

"We few, we happy few, we band of brothers; For he today that sheds his blood with me shall be my brother."
– William Shakespeare

What had been predicted as an easy triumph for the U.S. Marines - the swift conquest of a small remote island - quickly became a nightmare and one of the costliest battles of World War II. But the battle for Peleliu Island was not without its victories, for during this bloody campaign some of our social and military barriers were changed forever.

In the 1940's, racial segregation in the military was a fact of life. Although hard to understand today, it was part of our routine, and no one questioned it. Some, like me, a young kid from New Jersey, were hardly aware of it - until my experience at Peleliu.

After many weeks of intense fighting, we battle-weary Marines suffered heavy casualties with severe shortages of replacements and supplies.

On the fourth day of the battle I left the hospital ship after being treated for wounds, and was returning to my company ashore. The boat I took to shore, as well as the beach where we landed, was full of men and equipment to support the fighting. I asked around for the location of my unit, "Item" Company, Seventh Marines. A black truck driver pointed to the hills. I remembered vaguely that non-combatant African-American Marines volunteered wherever needed in combat or support, but this was the first time I had actually seen

African-American servicemen. I wondered who they were, and in which branch of the armed forces.

The dark-skinned men were working on the beach, stripped to their waists in the blistering tropical sun, transferring heavy ammunition from landing craft onto trucks for delivery to the front lines. This task was extremely dangerous at any time, but during battle, with enemy shells landing nearby, it was a heroic, thankless job that few of us wanted!

One of the men, the driver of a loaded ammunition truck, offered me a ride inland to the front lines. I accepted and climbed aboard. As the truck of explosive cargo bumped along the battle-scarred road, enemy shells crunched into the landscape nearby. I felt this selection of transportation was a dangerous choice on my part. Fortunately it was a short ride to the combat area, and I reached my unit safely. I thanked the truck driver, wished him well and quickly disembarked.

When I reached our company command post I located my top sergeant and reported for duty. He explained our battle assignment: "We are an under strength company, and our mission is to seize the ridge and mountains to our front. Gentlemen, get ready to earn your pay. That rocky, treeless mountain range is held by an elite, well-entrenched and hidden enemy who will defend this godforsaken place to their death."

The sweltering hot days and chilly cold nights added to the misery of this bloody assignment, which continued for weeks. Item Company's ranks grew thinner daily due to heavy casualties, rugged terrain and 115-degree heat.

Near a grim place called "Death Valley," our company's advance was halted. We were pinned down in a deadly crossfire by a concealed enemy supported by mortars and artillery. We suffered heavy casualties and urgently needed reinforcements. I was a company runner of messages and reported the new losses and the dangerous situation to the top sergeant at the command post. The top sergeant radioed for additional troops, medical corpsmen, water, ammunition and as many stretcher bearers as he could get.

The word came back from battalion headquarters: "Negative! No reinforcements, no stretcher bearers, no help or supplies for the present! All support and reserve units are committed in an all-out battle throughout the island."

With a grim determination born of desperation, the top sergeant turned over his command to the next in line and, summoning another Marine and me, set out to find help. Our trio jogged in the blazing sun to several rear headquarters command posts seeking assistance, but none was available.

When we reached the beach area, a young African-American overheard our situation, walked up and immediately volunteered his platoon's services. "We are from a Marine ammunition depot company and have had some infantry training," he told us.

The top sergeant looked at the dark-skinned sergeant in surprise. The races were so completely segregated during this era that we had no idea who these African-American servicemen were. Suddenly, I remembered seeing and talking to the African-American troops on the beach when I first returned to battle a week earlier. Now, for the first time, I realized they were Marines!

Our perplexed top sergeant tried to discourage the non-combatant volunteers from coming, stating they were not trained or qualified for the intensity of this battle. By now, the volunteers had heavily armed themselves and lined up behind their leader. I heard our seasoned professional Marine top sergeant say sharply "Well, don't say I didn't warn you." But I know he welcomed their aid.

We all returned to the battle area. There was carnage everywhere. The top sergeant reported to the acting company commander and said, “Sir, I have a platoon of black - I mean a platoon of Marine volunteers who came to help!”

The commanding officer said, “Thank God! Thank you all for coming. Sergeant, get our wounded to safety and our dead out.”

We watched in awe as the gallant volunteers did their job. While breaking through the surrounding enemy snipers, we saw more than one hold a casualty stretcher gently in one hand and, when necessary, fire an automatic weapon with the other hand.

One wounded Marine, probably the most bigoted man in our predominantly Southern unit, turned to me and said, “I’ll never put Negroes down again. These men are angels - black angels.”

The platoon of African-American volunteers made many dangerous trips to our company area for the wounded. With each return trip from the rear they brought badly needed ammunition, food, and precious water. It was nightfall when the evacuation of all the wounded was complete. Then the volunteers moved into empty foxholes and helped fight off a night skirmish.

Finally, Item Company’s Marines were relieved by a fresh U.S. Army infantry company. As the incoming soldiers passed what was left of our company on the road, the soldiers pointed at the African-American servicemen and hooted, taunting us, “Who are those guys in your outfit?”

Our senior sergeant bellowed, “Why, some of our best damn Marines, that’s who!”

This story originally appeared in “Chicken Soup for the Veteran’s Soul.”

A BRIDGE TOO FAR

"I'd rather be lucky than good!" – Words to live by

I always said I spent my entire career in the Marine Corps waiting for something to end - be it a night's duty, a week in the field, a month aboard ship, or a year in Okinawa. SCUBA school at Little Creek, Virginia was no different, and so it was only natural for our class to want to celebrate the end of yet another "journey."

There was a "Solid Gold" type topless bar not too far from the gate, and it was there the class decided to have an impromptu graduation party. Naturally we all had a good time, but there was one small problem - our actual graduation was not until the next morning. We had decided to have a "pre" graduation party because once the diplomas were handed out we would all be scattering to the four winds to return to our units, and it was unthinkable for us NOT to gather in celebration. That we would walk across the quarterdeck with hangovers and bloodshot eyes the next morning was a given, but we were willing to endure that because it was a small price to pay for a bit of camaraderie. But, of course, someone would somehow manage to screw the pooch despite all of our best intentions.

My dive buddy was a fellow I'll call 'Staff Sergeant Myers.' He was a good guy, and a solid Marine, who would one day end up being an instructor at the then nonexistent Marine Corps Combatant Diver School in Panama City. But on that particular night he was just another liberty risk.

As the night progressed we had noticed Myers was having a pretty good time, and was in fact getting drunk at the cyclic rate. It was soon obvious he would be unable to drive his vehicle back aboard the base, and as a group we decided to pull his car keys in order to avoid the possibility of an unfortunate incident. At least that was our intention. When we looked for our friend a bit later in the evening he was nowhere to be found, and a check of the parking lot revealed his car was gone. We didn't know what else to do, so we returned to our drinks until it was time to head home.

The next morning when the class turned to we discovered that Myers, who also happened to be the Class Commander, had not made it back the night before. A couple of us frantically began to call around to local hospitals in an attempt to locate our buddy, and when that didn't work we made the call everyone dreaded - to the police. Sure enough, Myers was in the can - charged with DUI.

It turned out that in his inebriated state he had mistaken the entrance to the Chesapeake Bay Bridge-Tunnel for the Little Creek Main Gate, and had pulled up to the toll booth expecting to be waved aboard the base. It was pretty obvious to the booth operator Myers was in no condition to drive, so he summoned a nearby cop who proceeded to take the hapless Marine into custody.

As anyone with any time in the Corps knows, a DUI, particularly for an officer or Staff NCO, is a career ender. As the saying goes, "To err is human, to forgive divine, but neither is Marine Corps policy." My friend had exercised some pretty bad judgment, and second chances are rare in the Corps. But Myers got one. For some reason his Colonel was in a charitable mood that day and didn't drop the hammer on him. That Colonel's decision was later rewarded as Staff Sergeant Myers went on to a long and distinguished career, eventually reaching the grade of Master Gunnery Sergeant. Sometimes it pays to be lucky!

HANGING THEM UP

By Michael E. Ruane

"The burdens of leadership are great. One of them is to be unpopular when necessary."

Very few Marines get to be Commandant, but even so anyone who has ever retired from the Corps, at any rank, will be able to identify with the way General Mundy must have felt when he retired:

On a clear, chilly morning in 1999, retired Marine Corps Commandant Carl E. Mundy Jr. stepped from a car outside the House of Representatives' Rayburn Office Building and marched up the steps for an appointment. It was 8 AM, a bit early for Capitol Hill. But he was calling on an old friend, Congressman John P. Murtha (D-Pa), who was himself a retired Marine Reserve officer and a member of a powerful House subcommittee. Mundy, who had taken off his general's stars and Commandant's laurel in 1995, used to come to the Hill attended by aides and advisers as a member of the Joint Chiefs of Staff and the commander of 170,000 Marines.

This morning, as the head of the USO, the venerable but haphazardly funded military morale agency, he was alone, in business attire, and had hat in hand. For fifteen minutes he waited at Murtha's office, until finally the congressman's scheduler arrived. There had been a mix-up. Murtha would not be in. "By the way," the scheduler politely asked, "Who are you?"

Since the time his doting father took him to see the newsreels of Marines fighting in the Pacific, and told the story about the Marines who saved him from a mugging in Philadelphia, it was preordained.

Mundy had tried to join up in high school in North Carolina, but was discouraged by his parents. He finally signed on with the reserves in college, hitchhiking home to tell his dismayed mother, who urged him, in vain, to take the paperwork back.

He'd worn his uniform to propose to his wife, served in Lebanon and the Philippines, directed battles in Vietnam, moved thirty-two times, reared two Marine sons, and was named Commandant by a president. He'd had the best job in the world, in the best outfit in the world. And few people had to ask who he was.

Then, one beautiful June evening, in a ceremony at the Iwo Jima Memorial, he found the Marine Band playing at his retirement, and it was all at an end. He bought a house on a cul-de-sac near Mount Vernon. He put his general's stars under glass in the living room, hung his ivory-handled sword in the den, and started building shelves for the basement.

A military career can be cruel that way. You finish up, hang out the flag, slap on a bumper sticker: "Semper Fi, Mac." And head for the fishing hole. No more uniforms. No obvious chain of command. For the first time in years, you pick a permanent place to live - Mundy picked one ten miles from the Pentagon. And if you were a high-ranking officer, you join the boards of corporations and foundations. Mundy joined seven.

But it doesn't replace who you were. Carl Mundy went through what scores of high-ranking military and civilian officials go through when they leave jobs in Washington, a town where the job defines the person - stepping off the public stage and plunging into the uncertain shadows of private life.

"It's almost like jumping out of an airplane," Mundy said. "You are weightless. There's no noise, or anything like that. The weight is off of you. You're not standing on your own feet. You're not bearing your own weight. You're just suspended in the air. You don't know whether it's time to go home and die in a couple of years, or whether you're going to devote yourself to your grandchildren, or whatever you're going to do."

So he decided to try starting over, becoming something else, someone else, agreeing to be Carl E. Mundy Jr., president and chief executive officer of the USO. Only for a while, though. After all, this was no childhood dream.

Mundy was born July 16, 1935, in Atlanta, the only child of a five-and-ten "set-up" man from South Carolina. His father's job was to establish a store, get it rolling and then, after about nine months, move on to another one. "I moved more as a dime store kid than I did as a Marine," Mundy said. After about ten years, his father changed jobs and the family settled in Waynesville, N.C.

In 1957, after officer training and four years at Auburn University, Mundy was graduated as a Marine Corps second lieutenant - "one bar was pinned on by my fourth-grade sweetheart, the other one by my mother," he said.

He shipped out in May 1958 to the Mediterranean and was one of five thousand Marines who assaulted the beaches of Lebanon in July, an operation designed to protect the government from a threatened coup and one he recalls as not the complete cakewalk it is often said to be.

Nine years later, he was sent to Vietnam - a captain, ripe for infantry company command, and finally bound for a real war. He was jubilant. Then, to his crushing disappointment, he was assigned to a staff job in DaNang. "You didn't want to go home and tell your kids that you were in DaNang the whole time," he said. Any Marine officer "worth a hoot" wanted to direct men in combat.

But Vietnam had plenty of that to go around, and in September 1967, near Con Thien, he got his wish. The Marines regularly battled it out in that area with disciplined North Vietnamese soldiers - "incredible people to fight," he noted.

"It was kind of a slug-it-out," he said. "There wasn't a heck of a lot of maneuvering. You kind of waited for them to come to you and fought, and then you went out to find them and then you fought. Nobody had brilliant strategy."

But combat, he added, was thrilling.

"In those periods of engagement, when he's shooting at you and you're shooting at him, or when you can see them or when they're coming at you... it is the most exhilarating moment of your life."

"It's a symphony. You can hear sporadic small-arms fire. That grows. When you know you're engaged is when you hear your machine guns start. And his machine guns pick up and by that time you realize, 'Okay, give me artillery.' And then you gotta see what's going on, and you go up on the ridge… and here they come," he said. "You almost think, 'Boy, I wish I had a camera.' It's beautiful."

The night before Mundy took the Commandant's job, on July 1, 1991, he and his wife went to the Iwo Jima Memorial, the Corps' shrine in Arlington. It was dusk and Mundy slowly walked around the thirty-foot bronze Marines - frozen like gods at the moment of the flag-raising on Mount Suribachi, and mounted atop a black granite base with the Corps' battles etched in gold. Belleau Wood. Meuse-Argonne. Wake Island. Bataan. Guadalcanal. Okinawa.

"Lord," Mundy prayed. "Don't let me do anything to screw up the Marine Corps." His petition was granted. But he stumbled often.

Although nominated by President Bush, he spent the bulk of his tour under President Clinton, a period of drastic military downsizing and turbulence over military social issues.

In 1993 Mundy, an opponent of "open homosexuality" in the service, was criticized for circulating a conservative group's anti-gay video among the Joint Chiefs of Staff. One angry editorial writer called him "this yahoo of a Marine officer."

Then in August of that year, he was rebuked when the Marines sought to ban the recruitment of married individuals to cut down on domestic discord in the Corps. The president was "astonished." Mundy apologized for "blind-siding" the administration.

Two months later he appeared on *60 Minutes* and told a national TV audience that Marine test scores showed minority officers did not shoot, swim or read compasses as well as white officers. he statement - taken out of context, he argued - drew outrage, and two days later Mundy apologized again. The lesson, he said, was "Never, ever go on *60 Minutes*."

But his four years also coincided with fiftieth-anniversary commemorations of World War II. Mundy got to attend ceremonies at Iwo Jima, Tarawa and Saipan, places he had watched flicker through the newsreels years before, with exotic names his father had mispronounced.

Four months after the commemoration at Iwo Jima, Carl Mundy's tour ended. On the evening of June 30, after he had formally handed over the Commandant's reins to his successor, Mundy, his family and friends gathered once again at the Iwo Jima Memorial, this time for his retirement ceremony.

As the band played his eldest son, Carl III, read a simple order signed by the Secretary of the Navy which Mundy memorized: "Effective midnight, 30 June, 1995, you are retired from active service in the Marine Corps. On that date you will have served forty-one years, nine months and twenty-three days as a United States Marine."

It's a long way from Iwo Jima to Bob Hope and the Dallas Cowboys cheerleaders, and the USO job was not one that Carl Mundy sought. The legendary organization, founded ten months before Pearl Harbor as the United Service Organizations for National Defense, has famously provided food, entertainment and moral support to service men and women during and in between all the nation's wars since 1941.

Supported only by public donations, its fortunes rose and fell, often sharply, with each conflict. Money poured in during wartime and dried up in times of peace.

After World War II, it almost shut down when the troops came home. It faced financial collapse in the mid-1980s. In the early 1990s, after a huge boost from the Persian Gulf War, it went into another nosedive, losing fifty thousand donors in about a year. There were fears it might not meet its payroll.

In 1995 the USO was still struggling. Its president, Chapman B. Cox, a former assistant secretary of the Navy and a Marine Corps lawyer during the Vietnam War, was stepping down after six years in the job.

"We needed somebody with stature," said Edward J. Christie Jr., a senior vice president at the USO who helped search for a replacement. "Someone who people outside these four walls would know."

Several weeks after he retired, Mundy had dinner with Cox, an old acquaintance, who asked him if he would be interested in the USO job. "I don't think so," Mundy replied. He'd had other offers. Virginia Military Institute wondered if he'd be interested in the top job there. He wasn't sure what he was going to do.

But Cox kept after him. Mundy visited USO headquarters at the Washington Navy Yard, in a building soon to be torn down, and found it "kind of dumpy." He met with the USO search committee and asked what they were looking for. Stability, organization, planning, they said... and fund-raising.

"Wrong guy," Mundy said he replied. "I can do all the front three for you... but I'm not a fund-raiser." He left thinking he was out of the running, but by the time he got home, after driving through the rain, the head of the search committee was on the phone offering him the job.

Mundy groaned. He hated begging for money. Plus, USO morale was poor. Focus was bad. Stability was questionable. Many people either didn't know what the USO was, or thought it had died years before. The Marine Corps it wasn't. Not by a long shot. But he still said yes. They faxed him a contract before he hung up.

Mundy has a terrific USO spiel. He tells the story of how, as a boy, he helped his mother spread mayonnaise on dozens of sandwiches, and how the whole town would go down to the USO center in the high school gym to give the GIs food, pencils and writing paper on Saturday night.

He tells how he waited in the rain for hours in Vietnam, water dripping off his helmet, to see a Bob Hope USO show in 1967, and how even today young service men and women constantly gravitate to USO centers in far-flung spots seeking a precious sliver of home.

In the three years after taking over he shook things up at the organization. He ended superfluous operations - things like the big, money-losing gift shop on Okinawa and the organization's scholarship program. He went temporarily into the red to acquire a $250,000 state-of-the-art computer system. He started an endowment, doubled the donor base and reduced staff turnover. Then he formally asked Congress - for the first time in USO history - for a government appropriation to permanently shield the USO from the vagaries of private funding.

Mundy cheerfully told everyone that he had the second best job in Washington - but the distance between Number One and Number Two seemed vast. Despite his diligence there was a sense of temporariness, a feeling that while he had set the organization on a sound course, and gave it his intellect and experience, that his heart was long ago taken by another. That was evident one day as he talked in his office, while two solitary Marines were practicing drill on a basketball court across from USO headquarters on Eberle Place. They were from the Marine barracks on I Street SE, where Mundy had once lived. They were practicing drill for a morning colors ceremony, a routine done with slow, careful steps and gold-handled swords. Mundy had always relished Marine ceremony. And he had promoted the sword as the symbol of the Corps.

But on that day, though his window looked out on the court below, he was busy in Suite 301 – a CEO in a big, bare office, across a narrow street and an unbridgeable divide from the life he had *really* loved.

This article appeared in the Washington Post on June 5, 1999

NOT A MARINE
But Still a Patriot

"A battle is won by those who are firmly resolved to win it." – Tolstoy

While we Marines take a lot of pride in being The Few and the Proud, and rightly so, we certainly are not the only patriots. Many fine Americans have chosen to serve our country in one of the other services, and we would be in trouble if they did not do so. Whenever I come across such a person I almost without exception give them the highest praise I can possibly give - that he could have, or should have, been a Marine. That is the case here. The following speech, delivered shortly after 9/11, is a wonderful example of the patriotic spirit that has made our great nation what it is. I take my hat off to Mr. Shul - he could have been a Marine!

"Thank you for the opportunity to address this rally today. It is not often that a fighter pilot is asked to be the keynote speaker. There is a rumor that they are unable to put two sentences together coherently. I'd like to dispel that rumor today by saying that I can do that, and in fact that I have written several books. I always wanted to be an author, and I ARE one now.

I'm a pretty lucky person really. I'm like the little boy who tells his father that when he grows up he wants to be a jet pilot, and his father replies, "Sorry son, you can't do both." I made that choice a long time ago and flew the jets. I was fortunate to live my dream, and then some. I survived something I shouldn't have, and today tell people that I am twenty-eight years old - as it has been that long since I was released from the hospital. It was like I received a second life, and in the past twenty-eight years I have gotten to see and do much, so much that I would not have thought possible. Returning to fly jets in the Air Force, flying the SR-71 on spy missions, spending a year with the Blue Angels, running my own photo studio... and so much more. And now, seeing our country attacked in such a heinous way.

Some of you here today have heard me speak before, and know that I enjoy sharing my aviation slide show. I have brought no slides to show you, as I feel compelled today to address different issues concerning this very difficult time in our nation's history.

I stand before you today, not as some famous person, or war hero. I am far from that. You know, they say a good landing is one you can walk away from, and a really great one is when you can use the airplane again. Well, I did neither... and I speak to you to today as simply a fellow American citizen.

Like you, I was horrified at the events of September 11th. But I was not totally surprised that such a thing could happen, or that there were people in the world who would perpetrate such deeds, willingly, against us. Having sat through many classified briefings while in the Air Force I was all too l aware of the threat, and I can assure you it has always been there in one form or another. And those of you who have served in the defense of this nation know all too well the response that is needed. In every fighter squadron I was in there was a saying that we knew to be true which said, 'when there was a true enemy, you negotiate with that enemy with your knee in his chest and your knife at his throat.'

Many people are unfamiliar with this way of thinking, and shrink from its ramifications. War is such a messy business, and there are many who want no part of it, but rush to bask in the security blanket of its victory.

I spent an entire military career fighting Communism, and was very proud to do so. We won that war, we beat one of the worst scourges to humankind the world has known. But it took a great effort, over many years of sustained vigilance and much sacrifice by so many whose names you will never know. And perhaps our nation, so weary from so long a cold war, relaxed too much and felt the world was a safer place with the demise of the Soviet Union. We indulged ourselves in our own lives, and gave little thought to the threats to our national security.

You know, normally my talks are laced with numerous jokes as I share my stories, but I have very few jokes to tell this afternoon. These murdering fanatics came into our land, lived amongst our people, flew on our planes, crashed them into our buildings, and killed thousands of our citizens. And nowhere along their gruesome path were they questioned or stopped. The joke is on us. We allowed this country to become soft.

We shouldn't really be too surprised this could happen. Did we really think that we could keep electing officials who put self above nation, and this would make us stronger? Did we really think that a strong economy adequately replaced a strong intelligence community? Did we imagine that a President who practically gave away the store on his watch was insuring national security? While our country was mired in the wasted excess of a White House sex scandal the drums of war beat loudly in foreign lands, and we were deaf. Our response was to give the man two terms in office, and even then barely half the American public exercised their right to vote. We have only ourselves to blame. Our elected officials are merely a reflection of our own values and what we deem important.

Did we not realize that America had become a laughing stock around the world? We had lost credibility, even amongst our allies. To our enemies we had no resolve. We made a lot of money, watched a lot of TV, and understood little about what was happening beyond our shores. We were, simply, an easy target.

But we are a country awakened now. We have been attacked in our homeland. We have now felt the reality of what an unstable and dangerous world it truly is. And still, in the face of this unprecedented carnage in our most prominent city, there are those who choose to take this opportunity to protest, and even burn the flag.

If I were the regents or alumni of certain large universities in this county, I would be embarrassed to be producing students of such ignorance and naïve notions. Like mindless sheep, they march with painted faces and trite sayings on signs, blissfully ignorant of the world they live in, and the system that protects them, hoping maybe to make the evening news. Perhaps if they had spent more time in class they would have learned that those who forget the past are condemned to repeat it. They might have learned that all it takes for evil to succeed in the world is for good people to stand by and do nothing. If they had simply gone back in history as recently as the Vietnam War they would have learned that an enemy that knows it can never defeat us militarily will persist as long as there is dissention and disruption in our land. Their ignorance can be understood, as their young empty minds have been filled with the re-written history tripe that tenured leftist professors can spew out with no fear of removal. But the unwitting aid they provide the enemy in disrupting the national resolve is unforgivable. I think this is wonderful country though, that gives everyone their voice of dissention. I am all for people expressing their views publicly because it makes it much easier for us to identify the truly foolish, and to know who cannot be counted on in times of crisis. These are the weak and cowardly who, when the enemy is crashing through the front door, will cower in the back room, counting on better men than themselves to make and keep them free. Well, the enemy is at our front door, and isn't it interesting that those who cry loudest and most often for their rights are usually those least willing to defend them.

I heard a student on TV the other day say that this war just wasn't in his plans and he would simply head to Canada if a draft occurred. Just wasn't in his plans. I wonder what plans the young men at the beaches of Normandy had that they never got to live. I wonder if it was in the plans of nineteen-year-old boys in Vietnam to lie dying in a jungle far from home. I guess the men and women at Pearl Harbor one morning had their plans slightly rearranged too. Gee, I hope we haven't inconvenienced this student. Those people in the World Trade Center have no more plans. It is up to us to have a plan now. And it isn't going to be easy. Who ever said it would? Just what part of our history spoke of how easy it was to form a free nation? It has never been easy and has always required vigilance and sacrifice, and sometimes war, to preserve this union. If it were easy, everyone would have done it. But no one else has, and we stand alone as the most unique country on earth.

And isn't it amazing that we have spent a generation stamping God out of our schools and government, and now as a nation have collectively turned to God in memorial services, prayer vigils and churches around this country.

I am also very disturbed to hear that there are people in this country at this particular time who feel it inappropriate to wear the flag on their lapel because they are on the news or in a public job, and that there are school officials who want to remove pro-American stickers so as not to offend foreign students. Well I am offended that these people call themselves Americans. I am offended that innocent people were killed in a mass attack of unthinkable proportions. And I am offended at listening to TV broadcasters speak to me condescendingly, with a bias that screams of their drowning in a cesspool of political correctness. And I pity the person who thinks they are going to remove this flag from *my* lapel!

This flag of ours is the symbol of all that is good about this country. America is an idea. It is an idea lived, and fought for, by a people. We are America, and this is our symbol. We are imperfect in many ways, but we continue to strive toward the ideal our forefathers laid down for us over 225 years ago. I could never imagine desecrating that symbol. Perhaps there are many people in this nation who have never been abroad, or in harm's way, and seen the flag upon their return. Those poor souls can never know the deep pride and honor one feels to see it wave, to know that there is still a good ol' USA. With all our warts we are still the greatest nation on earth, and the flag is the most powerful symbol of that greatness. When I was in grade school we used to say the Pledge of Allegiance every morning. It is something I never forgot. I wonder how many children even know that pledge today.

This flag is our history, our dreams, our accomplishments, indelibly expressed in bright red, white, and blue. This flag was carried in our Revolutionary War, although it had many less stars. But it persevered and evolved throughout a war we had no right to believe we could win. But we did, and built a country around it. This flag, tattered and battle worn, waved proudly from the mast as John Paul Jones showed the enemy what true resolve was. This banner was raised by the hands of brave men on a godforsaken island called Iwo Jima, and became a part of the most famous photo of the Twentieth Century. Those men are all dead now, but their legacy lives on in the Marine Memorial in Washington, DC. Those of you who have seen it will recall that inscribed within the stone monument are the words "Uncommon Valor Was a Common Virtue" - I don't believe you'll see the words, "it was easy" anywhere on it. This flag has even been to the moon, planted there for all time by men with a vision, and the courage to see it through.

I personally know what it is to see the flag, and feel something deep inside that makes you feel you are a part of something much bigger than yourself. I remember lying in a hospital bed, and I can vividly recall looking out the only window in the room and on Sundays seeing that big garrison flag flying proudly in the breeze. It filled the entire window, and filled my heart with a motivation that helped me leave that bed, and enabled me to be standing here today. And many years later, while fighting another terrorist over Libya, my backseater and I outraced Khaddafi's missiles in our SR-71 as we headed for the Mediterranean, and I can still clearly see that American flag patch on the shoulder of my space suit staring at me in the rear view mirror as we headed west, and it was a good feeling. Now don't ask me why we had rear view mirrors in the world's fastest jet. I can assure you, no one was gaining on us that day.

History will judge us. How we confront this chapter of American history will be important for the future of this great nation. This will be a war like none other we have endured. The combatants will not just be the soldier on the battlefront - it will also be fought by us, the citizens. We are on the battlefield now; the war has been brought to us. We will determine the outcome of this war by how well we remain vigilant, how patient we are with tightened security, how well we support the economy, and most importantly, in the resolve we show the enemy. There are some things worth fighting for, and this country is one of them.

I pray for our leaders at this time. In the Pacific during WW II Admiral Bull Halsey said, "There are no great men, just great circumstances, and how they handle those circumstances will determine the outcome of history." Our future and the future of coming generations are in our hands. Wars are not won just on military fronts, but by the resolve of the people. We must remain tenaciously strong in the pursuit of this enemy that threatens free people everywhere.

I am encouraged that we will win this war. Even before the first shot was finished being fired there were brave Americans on Flight 93 fighting back. These people were the first true heroes of this conflict, and gave their lives to save their fellow countrymen.

This nation, this melting pot of humanity, this free republic, must be preserved. This idea that is America is important enough to be defended. Fought for. Even died for. The enemy fears what you have, for if their people ever become liberated into a free society, tyrannical dictatorships would cease and they would lose power.

How can they ever understand this country of ours, so self-indulgent and diverse, yet when attacked, so united in the defense of its principals? This is the greatest country in the world because brave people sacrificed to make it that way. We are a collective mix of greatness and greed, hi-tech and heartland. We are the country of Mickey Mouse and Mickey Mantle; from John Smith and Pocahontas to John

Glenn and an Atlas booster; from Charles Lindbergh to Charley Brown; from Moby Dick to Microsoft; we are a nation that went from Kitty Hawk to Tranquility Base in less than seventy years; we are rock and roll, and the Bill of Rights; we are where everyone else wants to be, the greatest nation in the world.

The enemy does not understand the dichotomy of our society, but they should understand this; we will bandage our wounds, we will bury our dead; and then we will come for you… and we will destroy you and all you stand for.

I read this quote recently and would like to share it with you:

We are pressed on every side, but not crushed,
Perplexed, but not in despair,
Persecuted, but not abandoned,
Struck down, but not destroyed.

That is from II Corinthians. Not too long ago it would have been politically incorrect to quote from the Bible. I am so happy to be politically INCORRECT. And I am so proud to be an American.

Thank you all for coming out today and showing your support for your government, and your nation. You are the true patriots, you are the soldiers of this war, you are the strength of America."

Brian Shul served as a Vietnam era USAF fighter pilot with 212 combat missions. He was shot down near the end of the war and was so badly burned that he was given next to no chance to live. He did live, went on to fly SR-71s and completed a twenty year career in the Air Force. Has written four books on aviation and runs a photo studio. This is a speech he made in Chico California in the aftermath of the September 11th attack on the U.S.

THE KILLER ELITE

Evan Wright

"They are America's Warriors, and they are ready. These are United States Marines, and they are dangerous. They are poised, if and when the order comes, to wreak havoc on the enemies of their country." – Phil McCombs, Washington Post, Sept 13 2001

No matter how long you serve, or how many times you face tragedy, you never get completely insensitive to losing a friend. Not if you are human, anyway. As the war in Iraq progressed I of course followed developments there both on the news and through a number of military bulletin boards on the internet. It was while reading the posts on one of those sites that I came across the following article from Rolling Stone Magazine, and that is how I learned of the death of an old friend. Ed "Horsehead" Smith, a longtime Recon Marine and colleague, had put in his retirement papers before the war - but remained on active duty for the duration. It came as quite a shock to read of his demise. I was further saddened when I witnessed his death on a television news magazine a couple of weeks later - he was one of the Marines they had chosen to follow through the war, and it was surreal watching him be interviewed, and later medevaced. Ed was a good Marine and a decent man, and I include this story in his memory:

The invaders drove north through the Iraqi desert in a Humvee, eating candy, dipping tobacco and singing songs. Oil fires burn on the horizon, set during skirmishes between American forces and pockets of die-hard Iraqi soldiers. The four Marines crammed into this vehicle - among the very first American troops to cross the border into Iraq - are wired on a combination of caffeine, sleep deprivation, excitement and tedium. While watching for enemy fire and simultaneously belting out Avril Lavigne's "I'm With You," the driver, Corporal Joshua Ray Person, and the vehicle team leader, Sergeant Brad Colbert - both Afghan War veterans - have already reached a profound conclusion about this campaign: that the battlefield that is Iraq is filled with "fucking retards." There's the retard commander who took a wrong turn near the border, delaying the invasion by at least an hour. There's another officer, a classic retard, who has already begun chasing through the desert to pick up souvenirs thrown down by fleeing Iraqi soldiers: helmets, Republican Guard caps, rifles. There are the hopeless retards in the battalion-support sections who screwed up the radios and didn't bring enough batteries to operate the Marines' thermal-imaging devices. But in their eyes, one retard reigns supreme: Saddam Hussein - "We already kicked his ass once," says Person, spitting a thick stream of tobacco juice out his window. "Then we let him go, and he spends the next twelve years pissing us off even more. We don't want to be in this shit-hole country. We don't want to invade it. What a fucking retard."

The war began twenty-four hours ago as a series of explosions that rumbled across the Kuwaiti desert beginning at about six in the morning on March 20th. Marines sleeping in holes dug into the sand twenty miles south of the border with Iraq sat up and gazed into the empty expanse, their faces blank as they listened to the distant rumblings. There were 374 men camped out in the remote desert staging area, all members of the First Reconnaissance Battalion, which would lead the way during considerable portions of the invasion of Iraq, often operating behind enemy lines. These Marines had been eagerly anticipating this day since leaving their base at Camp Pendleton, California, more than six weeks before. Spirits couldn't have been higher. Later that first day, when a pair of Cobra helicopter gunships thumped overhead, flying north, presumably on their way to battle, Marines pumped their fists in the air and screamed, "Yeah! Get some!"

"Get some!" is the unofficial Marine Corps cheer. It's shouted when a brother Marine is struggling to beat his personal best in a fitness run. It punctuates stories told at night about getting laid in whorehouses in Thailand and Australia. It's the cry of exhilaration after firing a burst from a .50-caliber machine gun. "Get some!" expresses in two simple words the excitement, fear, feelings of power and the erotic-tinged thrill that come from confronting the extreme physical and emotional challenges posed by death, which is, of course, what war is all about. Nearly every Marine I've met is hoping this war with Iraq will be his chance to get some.

Marines call exaggerated displays of enthusiasm - from shouting "Get some!" to waving American flags to covering their bodies with Marine Corps tattoos - "moto." You won't ever catch Sergeant Brad Colbert, one of the most respected Marines in First Recon and the team leader I would spend the war with, engaging in any moto displays. They call Colbert the Iceman. Wiry and fair-haired, he makes sarcastic pronouncements in a nasal whine that sounds a lot like David Spade. Though he considers himself a "Marine Corps killer," he's also a nerd who listens to Barry Manilow, Air Supply and practically all the music of the 1980s except rap. He is passionate about gadgets - he collects vintage video-game consoles and wears a massive wristwatch that can only properly be "configured" by plugging it into his PC. He is the last guy you would picture at the tip of the spear of the invasion of Iraq.

The vast majority of the troops will get to Baghdad by swinging west onto a modern superhighway built by Hussein as a monument to himself and driving, largely unopposed, until they reach the outskirts of the Iraqi capital. Colbert's team will reach Baghdad by fighting its way through some of the crummiest, most treacherous parts of Iraq. Their job will be to screen the advance of a Marine battle force, the seven thousand-strong Regimental Combat Team One (RCT-1), through a 115-mile- long agricultural-and-urban corridor that runs between the cities of An Nasiriyah and Al Kut filled with thousands of well-armed fedayeen guerrilla fighters. Through much of this advance First Recon, mounted in a combination of seventy lightly armored and open-top Humvees and trucks, will race ahead of RCT 1, uncovering enemy positions and ambush points by literally driving right into them. After this phase of the operation is over, the unit will move west and continue its role as ambush hunters during the assault on Baghdad.

Reconnaissance Marines are considered among the best trained and toughest in the Corps. Major General James Mattis, commander of the Marine ground forces in Iraq, calls those in First Recon "cocky, arrogant bastards." They go through much of the same training as do Navy SEALS and Army Special Forces. They are physical prodigies who can run twelve miles loaded with 150-pound packs, then jump in the ocean and swim several more miles, still wearing their boots, fatigues and carrying their weapons and packs. They are trained to parachute, scuba dive, snowshoe, mountain climb and rappel from helicopters. Many of them are graduates of Survival Evasion Resistance Escape School, a secretive training facility where Recon Marines, fighter pilots, Navy SEALs and other military personnel in high-risk jobs are put through a simulated prisoner-of-war camp with student inmates locked in cages, beaten (within prescribed limits) and subjected to psychological torture overseen by military psychiatrists - all with the intent of training them to resist enemy captivity.

Paradoxically, despite all the combat courses Recon Marines are put through (it takes a couple of years for them to cycle through every required school), almost none are trained to drive Humvees and fight in them as a unit. Traditionally, their job is to sneak behind enemy lines in small teams, observe from afar and avoid contact with the enemy. What they are doing in Iraq - seeking out ambushes and fighting through them - is something they only started training for around Christmas, a month before being deployed to Kuwait. Corporal Person, the team's primary driver, doesn't even have a military operator's license for a Humvee and has only practiced driving in a convoy at night a handful of times.

General Mattis, who had other armored-reconnaissance units available to him - ones trained and equipped to fight through enemy ambushes in specialized, armored vehicles - says he choose First Recon for one of the most dangerous roles of the campaign because "what I look for in the people I want on the battlefield are not specific job titles but courage and initiative." By the time the war is declared over, Mattis will praise First Recon for having been "critical to the success of the entire campaign." The Recon Marines will face death nearly every day for a month and they will kill a lot of people, and a few of whose deaths Sergeant Colbert and his fellow Marines will no doubt think about and perhaps even regret for the rest of their lives.

Colbert's first impression of Iraq is that it looks like "fucking Tijuana." It's a few hours after his team's dawn crossing into Iraq. We are driving through a desert trash heap, periodically dotted with mud huts, small flocks of sheep and clusters of starved-looking, stick-figure cattle grazing on scrub brush. Once in a while you see wrecked vehicles: burnt-out car frames, perhaps left over from the first Gulf War, a wheel-less Toyota truck resting on its axles. Occasionally there are people, barefoot Iraqi men in robes. Some stand by the road, staring. A few wave. "Hey, it's ten in the morning!" says Person, yelling in the direction of one of the Iraqis we pass. "Don't you think you ought to change out of your pajamas?"

Person has a squarish head and blue eyes so wide apart his Marine buddies call him Hammerhead or Goldfish. He's from Nevada, Missouri, a small town where "NASCAR is sort of like a state religion." He speaks with an accent that's not quite Southern, just rural, and he was proudly raised working-poor by his mother. "We lived in a trailer for a few years on my grandpa's farm, and I'd get one pair of shoes a year from Wal-Mart." Person was a pudgy kid in high school who didn't play sports, was on the debate team and played any musical instrument - from guitar to saxophone to piano - he could get his hands on.

Becoming a Marine was a 180-degree turn for him. "I'd planned to go to Vanderbilt on a scholarship and study philosophy," he says. "But I had an epiphany one day. I wanted to do my life for a while, rather than think it." It often seems like the driving force behind this formerly pudgy, nonathletic kid's decision to enter the Corps and to join one of its most elite, macho units was so he could mock it, and everything around him. A few days before moving out of its desert camp in Kuwait to begin the invasion, his unit was handed letters sent by schoolchildren back home. Person opened one from a girl who wrote that she was praying for peace. "Hey, little tyke," Person shouted. "What does this say on my shirt? 'U.S. Marine!' I wasn't born on some hippie-faggot commune. I'm a death-dealing killer. In my free time I do push-ups until my knuckles bleed. Then I sharpen my knife."

As the convoy charges north into the desert, Person sings the Flock of Seagulls' song *I Ran (So Far Away)*. He says, "When I get out" - he's leaving the Marines in November - "I'm going to get a Flock of Seagulls haircut, then I'm going to become a rock star."

"Shut up, Person," Colbert says, peering intently at the dust-blown expanse, his M-4 rifle pointed out the window. Colbert and Person get along like an old married couple. Being a rank lower than Colbert, Person can never directly express anger to him, but on occasions when Colbert is too harsh and Person's feelings are hurt, the driving of the Humvee suddenly becomes erratic. There are sudden turns, and the brakes are hit for no reason. It will happen even in combat situations, with Colbert suddenly in the role of wooing his driver back with retractions and apologies. But generally, they seem to really like and respect each other. Colbert praises Person, whose job specialty is to keep the radios running - a surprisingly complex and vital job for the team - calling him "one of the best radio operators in Recon."

Obtaining Colbert's respect is no small feat. He maintains high standards of personal and professional conduct and expects the same from those around him. This year he was selected as team leader of the year in First Recon. Last year he was awarded a Navy Commendation for helping to take out an enemy missile battery in Afghanistan, where he led one of the first teams of Marines on the ground. Everything about him is neat, orderly and crisp. He grew up in an ultramodern house designed by his father, an architect. There was shag carpet in a conversation pit. One of his fondest memories, he tells me, was that before parties, his parents would let him prepare the carpet with a special rake. He is a walking encyclopedia of radio frequencies and encryption protocols and can tell you the exact details of just about any weapon in the U.S. or Iraqi arsenal. Colbert once nearly purchased a surplus British tank, and even arranged a loan through his credit union, but backed out only when he realized just parking it might run afoul of zoning laws in his home state, the "communist republic of California."

But there is another side to his personality. His back is a garish wash of heavy-metal tattoos. He pays nearly five thousand dollars a year in auto-motorcycle insurance due to outrageous speeding tickets; he routinely drives his Yamaha R1 racing bike at one hundred and thirty miles per hour. He admits to a deep-rooted but controlled rebellious streak that was responsible for his parents sending him to a military academy when he was in high school. His life, he says, is driven by a simple philosophy: "You don't want to ever show fear or back down, because you don't want to be embarrassed in front of the pack."

With Colbert located in the front passenger seat, providing security off the right side of the vehicle, left-side security is provided by Corporal Harold Trombley, a nineteen-year-old who mans the SAW machine gun in the rear passenger seat. Trombley is a thin, dark-haired and slightly pale kid from Farwell, Michigan. He speaks in a soft yet deeply resonant voice that doesn't quite fit his boyish face. One of his eyes is bright red from an infection caused by the continual dust storms. He has spent the past couple of days trying to hide it so he doesn't get pulled from the team.

Technically, he is a "paper Recon Marine," because he has not yet completed Basic Reconnaissance Course. He looks forward to combat as "one of those fantasy things you always hoped would really happen." In December, a month before his deployment, Trombley got married. He spends his idle moments writing down lists of possible names for the sons he hopes to have when he gets home. "It's up to me to carry on the Trombley name," he says.

Despite some of the other Marines' reservations about Trombley, Colbert feels he has the potential to be a good Marine. Colbert is always instructing him - teaching him how to use different communications equipment, how best to keep his gun clean. Trombley is an attentive pupil, almost a teacher's pet at times, and goes out of his way to quietly perform little favors for the entire team, like refilling everyone's canteens each day.

The other team member in the vehicle is Corporal Gabriel Garza, a twenty-one-year-old from Sebastian, Texas. He stands half out of the vehicle, his body extending from the waist up through a turret hatch. He mans the Mark-19 automatic grenade gun, the vehicle's most powerful weapon, mounted on top of the Humvee. His job is perhaps the team's most dangerous and demanding. Sometimes on his feet for as long as twenty hours at a time, he has to constantly scan the horizon for threats. Garza doesn't look it, but the other Marines credit him with being one of the strongest men in the battalion, and physical strength rates high among them. He modestly explains his reputation for uncanny strength by joking, "Yeah, I'm strong. I've got retard strength."

Colbert's team is part of a twenty-three-man platoon in Bravo Company. Along with First Recon's other two line companies - Alpha and Charlie - as well as its support units, the battalion's job is to hunt the desert for Iraqi armor, while other Marines seize oil fields to the east. During the first forty-eight hours of the invasion, Colbert's team finds no tanks and encounters hundreds of surrendering Iraqi soldiers - whom Colbert does his best to avoid, so as not to be saddled with the burden of searching, feeding and detaining them, which his unit is ill-equipped to do. Fleeing soldiers, some of them still carrying weapons, as well as groups of civilian families stream past Colbert's vehicle parked by a canal on his team's second night in Iraq. Colbert delivers instructions to Garza, who is keeping watch on the Mark-19: "Make sure you don't shoot the civilians. We are an invading army. We must be magnanimous."

"Magna-nous?" Garza asks. "What the fuck does that mean?"

"Lofty and kinglike," Colbert answers.

Garza considers this information. "Sure," he says. "I'm a nice guy." Colbert and Person often pass the time monitoring the sins committed by a Recon officer they nickname 'Captain America.' Colbert and other Marines in the unit accuse Captain America of leading the men on wild-goose chases, disguised as legitimate missions. Captain America is a likable enough guy. If he corners you, he'll talk your ear off about all the wild times he had in college, working as a bodyguard for rock bands such as U2, Depeche Mode and Duran Duran. His men feel he uses these stories as a pathetic attempt to impress them, and besides, half of them have never heard of Duran Duran.

Before First Recon's campaign is over, Captain America will lose control of his unit and be investigated for leading his men into committing war crimes against enemy prisoners of war. A battalion inquiry will clear him, but here in the field, some of his men fantasize about his death. "All it takes is one dumb guy in charge to ruin everything," says one. "Every time he steps out of the vehicle, I pray he gets shot."

Aside from Captain America's antics, there's an inescapable sense among Colbert's team that this is going to be a dull war. All that changes when they reach Nasiriyah on their third day in Iraq.

On March 23rd Colbert's team, in a convoy with the entire First Recon Battalion, cuts off from the backcountry desert trails and heads northwest to Nasiriyah, a city of about 300,000 on the Euphrates River. By late afternoon, the battalion becomes mired in a massive traffic jam of Marine vehicles about thirty kilometers south of the city. The Marines are given no word about what's happening ahead, though they get some clue when, before sundown, they begin to notice a steady flow of casualty-evacuation helicopters flying back and forth from Nasiriyah. Eventually, traffic grinds to a halt. The Marines turn off their engines and wait. During the past four days, no one on the team has slept for more than two hours a night, nor has anyone had a chance to remove his boots.

Everyone wears bulky chemical-warfare protection suits and carries gas masks. When they do sleep, in holes dug at each stop, they are required to keep their boots on and wear their protective suits. They live on MREs (meals ready to eat), which come in plastic bags about half the size of a phone book. Inside there are about half a dozen foil packets containing a meat or vegetarian I, such as meatloaf or pasta. More than half the calories in an MRE come from candy and junk food such as cheese pretzels and toaster pastries. Many Marines supplement this diet with massive amounts of freeze-dried coffee - they often just eat the crystals straight from the packet - chewing tobacco and over-the-counter stimulants including ephedra.

Colbert constantly harps on his men to drink water and to take naps whenever there is a chance, even questioning them on whether their pee is yellow or clear. When he comes back from taking a crap, Trombley turns the tables on him.

"Have a good dump, Sergeant?" he asks.

"Excellent," Colbert answers. "Shit my brains out. Not too hard, not too runny."

"That sucks when it's runny and you have to wipe fifty times," Trombley says conversationally.

"I'm not talking about that."Colbert assumes his stern teacher's voice. "If it's too hard or too soft, something's not right. You might have a problem."

Another big topic is music. Colbert attempts to ban any references to country music in his vehicle. He claims that the mere mention of country, which he deems "the Special Olympics of music," makes him physically ill. The Marines mock the fact that many of the tanks and Humvees stopped along the road are emblazoned with American flags or moto slogans such as "Angry American" or "Get Some." Person spots a Humvee with the 9/11 catchphrase "Let's Roll!" stenciled on the side.

"I hate that cheesy patriotic bullshit," Person says. He mentions Aaron Tippin's *Where the Stars and Stripes and Eagles Fly.* "Like how he sings those country white-trash images. 'Where eagles fly.' Fuck! They fly in Canada, too. Like they don't fly there? My mom tried to play me that song when I came home from Afghanistan. I was like, 'Fuck, no, Mom. I'm a Marine. I don't need to fly a little flag on my car to show I'm patriotic.'"

"That song is straight homosexual country music... Special Olympics-gay," Colbert says.

Colbert's team spends the night by the highway. Late in the night, we hear artillery booming up ahead in the direction of Nasiriyah. The ground trembles as a column of massive M1A1 tanks rolls past, a few feet from where the Marines are resting.

A couple of hours after sunrise on the 24th they tune in to the BBC on a shortwave radio that Colbert carries in the Humvee and hear the first word of fighting up the road in Nasiriyah. A while later Colbert's platoon commander, Lieutenant Nathan Fick, holds a briefing for the three other team leaders in the twenty-three-man platoon. Fick, who's twenty-five, has the pleasant good looks of a former altar boy, which he is. The son of a successful Baltimore attorney, he went through Officer Candidate School after graduating from Dartmouth. This is his second deployment in a war. He commanded a Marine infantry platoon in Afghanistan. But like Colbert and the six other Marines in the platoon who also served in Afghanistan, he saw very little shooting.

Fick tells his men that the Marines have been taking heavy casualties in Nasiriyah. Yesterday, the town was declared secure. But then an Army supply unit traveling near the city came under attack from an Iraqi guerrilla unit of Saddam Hussein loyalists called 'fedayeen.' These fighters, Fick says, wear civilian clothes and set up positions in the city among the general populace, firing mortars, rocket-propelled grenades (RPGs) and machine guns from rooftops, apartments and alleys. They killed or captured twelve soldiers from the Army supply unit, including a woman. Overnight, a Marine combat team from Task Force Tarawa attempted to move into the city across the main bridge over the Euphrates. Nine Marines lost their lives, and seventy more were injured.

First Recon has been ordered to the bridge to support Task Force Tarawa, which barely controls its southern approach. Fick can't tell his men exactly what they're going to do when they get to the bridge, as the plans are still being drawn up at a higher level. What he does tell the men is that their rules of engagement have changed. Until now they've let armed Iraqis pass, sometimes even handing them food rations. Now, Fick says, "Anyone with a weapon is declared hostile. If it's a woman walking away from you with a weapon on her back, shoot her."

At 1:30 PM the 374 Marines of First Recon form up on the road and start rolling north toward the city. Given the news of heavy casualties during the past twenty-four hours, it's a reasonable assessment that everyone in the vehicle has a better than average chance of getting killed or injured in Nasiriyah.

The air is heavy with a fine, powdery dust that hangs like dense fog. Cobras clatter directly overhead, swooping low with the grace of flying sledgehammers. They circle First Recon's convoy, nosing down through the barren scrubland on either side of the road, hunting for enemy shooters. Before long, we are on our own. The helicopters are called off because fuel is short. The bulk of the Marine convoy is held back until the Iraqi forces ahead are put down. One of the last Marines we see standing by the road pumps his fist as Colbert's vehicle drives past and shouts, "Get some!"

We drive into a no man's land. A burning fuel depot spews fire and smoke. Garbage is strewn on either side of the road as far as the eye can see. The convoy slows to a crawl, and the Humvee fills with a black cloud of flies.

"Now, this looks like Tijuana," says Person.

"And this time I get to do what I've always wanted to do in T.J.," Colbert answers. "Burn it to the ground."

There is a series of thunderous, tooth-rattling explosions directly to the vehicle's right. We are even with a Marine heavy-artillery battery set up next to the road, firing into Nasiriyah, a few kilometers ahead. There's a mangled Humvee in the road. The windshield is riddled with bullet holes. Nearby are the twisted hulks of U.S. military-transport trucks, then a blown-up Marine armored vehicle. Marine rucksacks are scattered on the road, clothes and bedrolls spilling out.

We pass a succession of desiccated farmsteads - crude, square huts made of mud, with starving livestock in front. The locals sit outside like spectators. A woman walks past with a basket on her head, oblivious to the explosions. No one has spoken for ten minutes, and Person cannot repress the urge to make a goofy remark. He turns to Colbert, smiling. "Hey, you think I have enough driving hours now to get my Humvee license?"

We reach the bridge over the Euphrates. It is a long, broad concrete structure. It spans nearly a kilometer and arches up gracefully toward the middle. On the opposite bank, we glimpse Nasiriyah. The front of the city is a jumble of irregularly shaped two and three-story structures. Through the haze, the buildings appear as a series of dim, slanted outlines, like a row of crooked tombstones.

Nasiriyah is the gateway to ancient Mesopotamia, the Fertile Crescent lying between the Euphrates, just above us, and the Tigris, a hundred kilometers north. This land has been continuously inhabited for five thousand years. It was here that humankind first invented the wheel, the written word and algebra. Some scholars believe that Mesopotamia was the site of the Garden of Eden. After three days in the desert, the Marines are amazed to find themselves in this pocket of tropical vegetation. There are lush groves of palm trees all around, as well as fields where tall grasses are growing. As Marine artillery rounds explode around us, Colbert keeps repeating, "Look at these fucking trees."

While two First Recon companies are instructed to set up positions on the banks of the Euphrates, Bravo Company waits at the foot of the bridge, about two hundred meters away from the river's edge. No sooner are we settled than machine-gun fire begins to rake the area. Incoming rounds make a zinging sound, just like they do in Bugs Bunny cartoons. They hit palm trees nearby, shredding the fronds, sending puffs of smoke off the trunks. Marines from Task Force Tarawa to our right and to our left open up with machine guns. First Recon's Alpha and Charlie companies begin blasting targets in the city with their heavy guns.

Enemy mortars start to explode on both sides of Colbert's vehicle, about a hundred and fifty meters distant. "Stand by for shit to get stupid," Person says, sounding merely annoyed. He adds, "You know that feeling before a debate when you gotta piss and you've got that weird feeling in your stomach, then you go in and kick ass?" He smiles. "I don't have that feeling now."

Marine helicopters fly low over a palm grove across the street, firing rockets and machine guns. It looks like we've driven into a Vietnam War movie. As if on cue, Person starts singing a Creedence Clearwater Revival song. This war will need its own theme music, he tells me. "That fag Justin Timberlake will make a soundtrack for it," he says, adding with disgust, "I just read that all these pussy faggot pop stars like Justin Timberlake and Britney Spears were going to make an anti-war song. When I become a pop star, I'm just going to make pro-war songs."

While Person talks, there's a massive explosion nearby. An errant Marine artillery round hits a power line and detonates overhead, sending shrapnel into a vehicle ahead of ours. A group of six Marines is also hit. Two are killed immediately; the four others are injured. Through the smoke, we can hear them screaming for a medic. Everyone takes cover in the dirt. I lie as flat on the ground as possible. I look up and see a Marine cursing and wiggling, trying to pull down his chemical-protection suit. The pants don't have zippers in the front. You have to unhook suspenders and wrestle them down, which is especially tough when you're lying sideways. It's a Marine in Colbert's platoon, one of his closest friends, Sergeant Antonio Espera.

Espera grew up in Riverside, California, and was, by his own account, truly a "bad motherfucker" - participating in all the violent pastimes available to a young Latino from a broken home and raised partially in state facilities. With his shaved head and deep-set eyes, he's one of the scariest-looking Marines in the platoon, but Espera makes no show of trying to laugh off his fear.

The guy on my other side is another Bravo team leader, Sergeant Larry Sean Patrick. He's looked up to about as much as Colbert is. I ask him what the hell we're doing just waiting around while the bombs fall. His response is sobering. He tells me the platoon is about to be sent on a suicide mission. "Our job is to kamikaze into the city and collect casualties," he says.

"How many casualties are there?" I ask.

"Casualties?" he says. "They're not there yet. We're the reaction force for an attack that's coming across the bridge. We go in during the fight to pick up the wounded."

I don't know why, but the idea of waiting around for casualties that don't exist yet strikes me as more macabre than the idea of actual casualties. Yet despite how much it sucks here - by this bridge, taking heavy fire - it's kind of exciting, too. I had almost looked down on the Marines' shows of moto, the way they shouted "get some" and acted all excited about being in a fight. But the fact is, there's a definite sense of exhilaration every time there's an explosion and you're still there afterward. There's another kind of exhilaration, too. Everyone is side by side facing the same big fear: death. Usually, death is pushed to the fringes of things you do in the civilian world. Most people face their end pretty much alone, with a few family members if they are lucky. Here, the Marines face death together, in their youth. If anyone dies, he will do so surrounded by the very best friends he believes he will ever have.

As mortars continue to explode around us, I watch Garza pick through an MRE. He takes out a packet of Charms candies and hurls it into the gunfire. Marines view Charms as almost infernal talismans. A few days earlier, in the Humvee, Garza saw me pull Charms out of my MRE pack. His eyes lit up and he offered me a highly prized bag of cheese pretzels for my candies. He didn't explain why. I thought he just really liked Charms until he threw the pack he'd just traded me for out the window. "We don't allow Charms anywhere in our Humvee," Person said, in a rare show of absolute seriousness. "That's right," Colbert said, cinching it. "They're fucking bad luck."

A fresh pair of Marine gunships flies overhead, firing rockets into a nearby grove of palm trees. Bravo Marines leap up after one of the helicopters fires a TOW missile that sends up a large orange fireball from the trees. "Get some!" the Marines shout.

For nearly six hours we are pinned down, waiting, we think, to storm into Nasiriyah. But after sunset plans are changed, and First Recon is called back from the bridge to a position four kilometers into the trash-strewn wastelands south of the city. When the convoy stops in relative safety, away from the bridge, Marines wander out of the vehicles in high spirits. First Recon's Alpha Company killed at least ten Iraqis across the river from our position. They come up to Colbert's vehicle to regale his team with exploits of their slaughter, bragging about one kill in particular, a fat fedayeen in a bright-orange shirt. "We shredded him with our .50-cals," one says.

It's not just bragging. When Marines talk about the violence they wreak, there's an almost giddy shame, an uneasy exultation in having committed society's ultimate taboo, and doing it with state sanction.

"Well, good on you," Colbert says to his friend.

The Recon Marines are told they will be pushing north through Nasiriyah at dawn, along a route they've deemed "sniper alley." At midnight, Espera and I share a last cigarette. We climb under a Humvee for cover and lie on our backs, passing it back and forth.

Past midnight, Marine artillery booms into the city. Back in the Humvee, Trombley once again talks about his hopes of having a son with his new young bride when he returns home.

"Never have kids, Corporal," Colbert lectures. "One kid will cost you $300,000. You should never have gotten married. It's always a mistake." Colbert often proclaims the futility of marriage. "Women will always cost you money, but marriage is the most expensive way to go. If you want to pay for it, Trombley, go to Australia. For a hundred bucks, you can order a whore over the phone. Half an hour later, she arrives at your door, fresh and hot, like a pizza."

Despite his bitter proclamations about women, if you catch Colbert during an unguarded moment, he'll admit that he once loved one girl who jilted him, a junior-high-school sweetheart whom he dated on and off for ten years and was even engaged to until she left him to marry one of his closest buddies. "And we're still all friends," he says, sounding almost mad about it. "They're one of those couples that likes to takes pictures of themselves doing all the fun things they do and hang them up all over their goddamn house. Sometimes I just go over there and look at the pictures of my ex-fiancée doing all those fun things I used to do with her. It's nice having friends."

Just after sunrise, First Recon's seventy-vehicle convoy rolls over the bridge on the Euphrates and enters Nasiriyah. It's one of those sprawling Third World mud-brick-and-cinder-block cities that probably looks pretty badly rubbled even on a good day. This morning, smoke curls from collapsed structures. Most buildings facing the road are pockmarked and cratered. Cobras fly overhead spitting machine-gun fire. Dogs roam the ruins.

The convoy stops to pick up a Marine from another unit who is wounded in the leg. A few vehicles come under machine-gun and RPG fire. The Recon Marines return fire and redecorate an apartment building with about a dozen grenades fired from a Mark-19. In an hour, we clear the outer limits of the city and start to head north. Dead bodies are scattered along the edges of the road. Most are men, enemy fighters, some with weapons still in their hands. The Marines nickname one corpse Tomato Man, because from a distance he looks like a smashed crate of tomatoes in the road. There are shot-up cars and trucks with bodies hanging over the edges. We pass a bus, smashed and burned, with charred human remains sitting upright in some windows.

We drive on, pausing a few kilometers ahead for the battalion to call in an air strike on an Iraqi armored vehicle up the road. Next to me, Trombley opens up an MRE and furtively pulls out a pack of Charms. "Keep it a secret," he says. He unwraps the candies and stuffs them into his mouth.

At ten in the morning first Recon is ordered off Highway 7, the main road heading north out of Nasiriyah, and onto a narrow dirt trail, to guard the main Marine fighting force's flanks. There's a dead man lying in a ditch where we turn off the highway. Two hundred meters past the corpse there's a farmhouse with a family out front, waving as we drive by. At the next house, two old ladies in black jump up and down, whooping and clapping. A bunch of bearded men shout, "Good! Good! Good!" The Marines wave back. In the span of a few minutes they have gone from kill-anyone-that-looks-dangerous mode to smiling and waving as if they're on a float in the Rose Bowl parade.

"Stay frosty, gents," Colbert warns. "No matter what you see, we're in backcountry now, and we're all alone."

The road has dwindled down to a single narrow lane. We crawl along at a couple of miles per hour. There are farmhouses every few hundred meters. The Marines stop and toss bright-yellow humanitarian food packages at clusters of civilians. As kids run out to grab them, Colbert waves: "You're welcome. Vote Republican." He gazes at the "ankle biters" running after the food rations and says, "I really thank God I was born American. I mean, seriously, it's something I lose sleep over." The demeanor of the civilians we pass has suddenly changed. They've stopped waving. Many avoid eye contact with us altogether. Over the radio, we hear that RCT-1 is in contact with enemy forces at a town a few kilometers to the north. As we continue along the road, we begin to notice that villagers on the other bank of the canal are fleeing in the opposite direction. Two villagers approach a Humvee behind Colbert's and warn the Marines through hand gestures that something bad lies ahead. The convoy stops.

We are at a bend in the road, with a five-foot-high berm to the left. Shots are fired directly ahead of us.

"Incoming rounds," Person announces.

"Damn it," says Colbert. "I have to take a shit."

Instead Colbert picks up a 203 round, kisses its nose, and slides it into the lower chamber of his weapon. He opens the door and climbs up the embankment to observe a small cluster of homes on the other side. He signals for all the Marines to come out of the vehicle and join him on the berm. Marines from another platoon fire into the hamlet with rifles, machine guns and Mark-19s. But Colbert does not clear his team to fire. He can't discern any targets. About two kilometers up the road, where First Recon's Alpha Company is stopped, suspected fedayeen open up with machine guns and mortars. Alpha takes no casualties. The battalion calls in an artillery strike on the fedayeen positions.

The team gets back in the Humvee. Trombley sits in the back seat eating spaghetti directly out of a foil MRE pack, squeezing it into his mouth from a hole in the corner. "I almost shot that man," he says, referring to a farmer in the hamlet on the other side of the berm.

"Not yet," Colbert says. "Put your weapon on safe."

Nobody speaks for a solid ten minutes. A vicious sandstorm is kicking up. Fifty- to sixty-mile-per-hour winds buffet the side of the vehicle. Visibility drops, and the air fills with yellow dust. The battalion is hemmed in on narrow back roads with enemy shooters in the vicinity.

RCT-1 is now waiting outside a town about six kilometers ahead. Its commander has reported taking fire from the town, and First Recon plans to bypass it. Colbert explains the situation to his men.

"Why can't we just go through the town?" Trombley asks.

"I think we'd get smoked," Colbert says.

Fifteen minutes later, we start moving north. Everyone in Colbert's vehicle believes we are taking a route that bypasses the hostile town, Al Gharraf. Then word comes over the radio of a change in plan. We are driving straight through.

Colbert's vehicle comes alongside the walls of the town, which looks like a smaller version of Nasiriyah. The street we are on, now paved, bears left. As Person makes the turn, the wall of a house directly to my right and no more than three meters from my window erupts with muzzle flashes and the clatter of machine-gun fire. The vehicle takes twenty-two bullets, five of them in my door. The light armor that covers much of the Humvee (eighth-inch steel plates riveted over the doors) stops most of them, but the windows are open and there are gaps in the armor. A bullet flies past Colbert's head and smacks into the frame behind Person's. Another round comes partially through my door.

We have barely entered the city, and it's a two-kilometer drive through it. Ahead of us, a Bravo Marine driving in an open Humvee takes a bullet in his arm.

The shooting continues on both sides. Less than half an hour before, Colbert had been talking about stress reactions in combat. In addition to the embarrassing losses of bodily control that twenty-five percent of all soldiers experience, other symptoms include time dilation, i.e., time slowing down or speeding up; vividness, a starkly heightened awareness of detail; random thoughts, the mind fixating on unimportant sequences; memory loss; and, of course, your basic feelings of sheer terror. In my case, hearing and sight become almost disconnected. I see more muzzle flashes next to the vehicle but don't hear them. In the seat beside me, Trombley fires three hundred rounds from his machine gun. Ordinarily, if someone was firing a machine gun that close to you, it would be deafening. His gun seems to whisper.

The look on Colbert's face is almost serene. He's hunched over his weapon, leaning out the window, intently studying the walls of the buildings, firing bursts from his M-4 and grenades from the 203 tube underneath the main barrel. I watch him pump in a fresh grenade, and I think, "I bet Colbert's really happy to be finally shooting a 203 round in combat." I remember him kissing the grenade earlier. Random thoughts. I study Person's face for signs of panic, fear or death. My fear is he'll get shot or freak out, and we'll get stuck on this street. But Person seems fine. He's slouched over the wheel, looking through the windshield, an almost blank expression on his face. The only thing different about him is he's not babbling his opinions on Justin Timberlake or some other pussy faggot retard who bothers him.

Trombley pauses from shooting out his window and turns around with a triumphant grin. "I got one, Sergeant!" he shouts.

Colbert ignores him. Trombley eagerly goes back to shooting at people out his window. A gray object zooms toward the windshield and smacks into the roof. The Humvee fills with a metal-on-metal scraping sound, which I do hear. Earlier that day Colbert had traded out Garza for a Mark-19 gunner from a different unit. The guy's name is Corporal Walt Hasser. Hasser's legs twist sideways. A steel cable has fallen or been dropped over the vehicle. Another one falls on it and scrapes across the roof.

Colbert calls out, "Walt, are you OK?" There's silence. Person turns around, taking his foot off the gas pedal.

The vehicle slows and then wanders slightly to the left. "Walt?" Person calls.

"I'm OK!" he says, sounding almost cheerful. Person has lost his focus on moving the vehicle forward. We slow to a crawl. Person later says that he was worried one of the cables dropped on the vehicle might have been caught on Hasser. He didn't want to accelerate and somehow leave him hanging from a light pole by his neck in downtown Gharraf.

"Drive, Person!" Colbert shouts.

Person picks up the pace, and there is silence outside. We are still in the town, but no one seems to be shooting at us.

"Holy shit! Did you see that? We got fucking lit up!" Colbert is beside himself, laughing and shaking his head. "Holy shit!"

"Before we start congratulating ourselves," Person says, "we're not out of this yet."

We pass a mangled, burned car on the right, then Person makes a left into more gunfire. Set back from the road are several squat cinder-block buildings, like an industrial district. I see what looks like white puffs of smoke streaking out from them: more enemy fire. Person floors the Humvee. Colbert and Trombley start shooting again.

"I got another one!" Trombley shouts.

There's a white haze in the distance: the end of the city. We fly out onto a sandy field that looks almost like a beach. There's so much sand blowing in the air - winds are still at about sixty miles per hour - it's tough to see anything. There's gunfire all around. The Humvee drives about twenty meters into the sand, then sinks into it. Person floors the engine, and the wheels spin. The Humvee has sunk up to the door frames in tar. It's a sobka field. Sobka is a geological phenomenon peculiar to the Middle East. It looks like desert on top, with a hard crust of sand an inch or so thick, which a man could possibly walk on, but break through the crust and beneath it's the La Brea tar pits, quicksand made of tar.

Colbert jumps out and runs to the other Recon vehicles, lined up now, shooting into the city. He runs down the lines of guns, shouting, "Cease fire! Assess the situation!"

Back at Colbert's Humvee, one of his superiors pounds on the roof and shouts, "Abandon the Humvee!" He adds, "Thermite the radios!" He is referring to a kind of intense-heat grenade used to destroy sensitive military equipment before abandoning it.

Colbert jumps up behind him. "Fuck, no! I'm not thermiting anything. We're driving this out of here!"

He dives under the wheel wells with bolt cutters, slicing away the steel cables, a gift of the defenders of Gharraf, wrapped around the axle. A five-ton support truck backs up, its driver taking fire, and Marines attach towing cables to our axle. Within half an hour, Colbert's vehicle is freed and limping to Recon's camp, a few kilometers distant, for the night.

The Bravo Marines spend half an hour recounting every moment of the ambush. Aside from the driver in the other platoon who was shot in the arm, no one was hit. They laugh uproariously about all the buildings they blew up. Privately, Colbert confesses to me that he had absolutely no feelings going through the city. He almost seems disturbed by this. "It was just like training," he says. "I just loaded and fired my weapon from muscle memory. I wasn't even aware what my hands were doing."

That night we are rewarded with the worst sandstorm we have experienced in Iraq. Under a pitch-black sky, sand and pebbles kicked up by sixty-mile-per-hour winds pelt sleeping bags like hail. Then it rains. Lightning flashes intermingle with Marine artillery rounds sailing into the city. Just before turning in, I smell a sickly-sweet odor. During chemical-weapons training before the war, we were taught that some nerve agents emit unusual, fragrant odors. I put on my gas mask and sit in the dark Humvee for twenty minutes before

Person tells me what I'm smelling is a cheap Swisher Sweet cigar that Espera is smoking underneath his Humvee.

The next morning at dawn, Lieutenant Fick tells his Marines, "The good news is, we will be rolling with a lot of ass today. RCT-1 will be in front of us for most of the day. The bad news is, we're going through four more towns like the one we hit yesterday."

We roll on. We pass dead bodies in the road again, men with weapons by their sides, then more than a dozen trucks and cars burned and smoking by the road. Many have a burned corpse or two of Iraqi soldiers who died after crawling five or ten meters away from the vehicle before they expired, hands still grasping forward on the pavement. Just north of here, at another stop, Marines in Fick's vehicle machine-gun four men in a field who appear to be stalking us. It's no big deal. Since the shooting started in Nasiriyah forty-eight hours ago, firing weapons and seeing dead people has become almost routine.

First Recon is next sent several kilometers up the road to the edge of another town, Ar Rifa. Colbert's team stops thirty meters from the town's outer walls. The winds have died down, but dust is so thick in the air that it looks like twilight at noon. An electrical substation is on fire next to Colbert's vehicle, adding its own acrid smoke. Shots come from the town, and Colbert's team fires back.

After night falls outside of Rifa, another bad day in Iraq ends with a new twist: a friendly-fire incident. A U.S. military convoy moving up the road in complete darkness mistakenly opens fire on First Recon's vehicles. Inside his Humvee, Sergeant Colbert sees the "friendly" red tracer rounds coming from the approaching convoy and orders everyone down. One round slices through the rear of the Humvee, behind the seat where Trombley and I are sitting.

Later, we find out from Fick that we were shot up by Navy Reservist surgeons on their way to set up a mobile shock-trauma unit on the road ahead. "Those were fucking doctors who do nose and tit jobs," Fick tells the men.

A half-hour after the friendly-fire incident, First Recon is ordered to immediately drive forty kilometers through back roads to the Qal'at Sukkar airfield, deep behind enemy lines. "Well, I guess we won't be sleeping tonight," Colbert says.

The drive takes about three hours. On the way, the men are informed that they will be setting up an observation post on the field to prepare for a parachute assault that British forces are going to execute at dawn. But plans change again at sunrise. At 6:20 AM, after the Bravo Marines have slept for about ninety minutes, Colbert is awakened and told his men have ten minutes to race onto the airfield, six kilometers away, and assault it. At 6:28, Colbert's team is in the Humvee driving with thirty other Recon vehicles down a road they've never even studied on a map. They're told over the radio they will face enemy tanks.

"Everything and everyone on the airfield is hostile," Colbert says, passing on a direct order from his commander.

Next to me in the rear seat, Trombley says, "I see men running."

"Are they armed?" Colbert asks.

"There's something," Trombley says.

I look out Trombley's window and see a bunch of camels.

"Everyone's declared hostile," Colbert says. "Light them up."

Trombley fires a burst or two from his SAW.

The Humvees race onto the airfield and discover it's abandoned, nothing but crater-pocked airstrips. Nevertheless, they've beaten the British to it. The landing is called off.

"Gentlemen, we just seized an airfield," Colbert says. "That was pretty ninja."

An hour later, the Marines have set up a camp off the edge of the airfield. They are told they will stay here for a day or longer. This morning, the sun shines and there's no dust in the air. For the first time in a week, many of the Marines take their boots and socks off. They unfurl camo nets for shade and lounge beside their Humvees. A couple of Recon Marines walk over to Trombley and tease him about shooting camels.

"I think I got one of those Iraqis, too. I saw him go down."

"Yeah, but you killed a camel, too, and wounded another one."

The Marines seem to have touched a nerve.

"I didn't mean to," Trombley says defensively. "They're innocent."

A couple of hours later, the word came in. Horsehead is dead. The beloved former First Sergeant in the Marine First Reconnaissance Battalion, a powerfully built 230-pound African-American named Edward Smith, was felled by an enemy mortar or artillery blast while riding atop an armored vehicle outside Baghdad on April 4th. He died in a military hospital the next day. Horsehead, thirty-eight, had transferred out of First Recon to an infantry unit before the war started. News of his death hits the Recon battalion hard. Sergeant Rudy Reyes is one of the first to hear of it. He moves along the camp's perimeter just outside Baghdad, spreading the word. "Hey, brother," he says softly, "I just came by to tell you Horsehead died last night."

Now, a couple of days later, following a brief sundown memorial around an M-4 rifle planted upright in the dirt in honor of their fallen comrade - Marines in Bravo Company's Second Platoon gather under their camouflage nets trading

Horsehead stories. Reyes repeats a phrase Horsehead always used back home at Camp Pendleton in San Diego. Before loaning anyone his truck, which had an extensive sound-equalizer system, he'd say, "You can drive my truck. But don't fuck with my volumes." For some reason, repeating the phrase makes Reyes laugh almost to the verge of tears.

It's April 8th. Army and Marine units began their final assault on Baghdad several hours ago. First Recon, however, will not be heading into the Iraqi capital just yet. It's feared that Iraqi Republican Guard units may be massing for a counterattack in a town called Ba'qubah, fifty kilometers north of Baghdad. First Recon receives orders to head north and attack these forces. Sergeant Colbert, whose team I am riding with, and the rest of the Marines stop reminiscing about Horsehead and load their Humvees.

About two hundred Recon Marines are slated for this mission. If the worst-case fears of their commanders are true, they will be confronting several thousand Iraqis in tanks. In the best-case scenario, they will merely be assaulting through about thirty kilometers of known ambush points along the route to Ba'qubah. "Once again, we will be at the absolute tippity-tip of the spear, going into the unknown," says Lieutenant Nathaniel Fick, briefing his men just before the mission. Most of the Marines are in high spirits. "It beats sitting around doing nothing while everybody else gets to have fun attacking Baghdad," says Corporal Person before taking his position in the driver's seat of Colbert's Humvee. Colbert, however, just stares out his window at the fading light and mumbles something I can't quite make out. I ask him to repeat it, and he waves it off. "It was nothing," he says. "I was just thinking about Horsehead."

Taking the lead of First Recon's fifty-vehicle column, Colbert's Humvee drives out past the camp's concertina wire and into the eastern outskirts of Baghdad. We pass newly liberated Iraqis in the throes of celebration. Though the city center will not fall for another twenty-four hours, freedom fills the air, along with the stench of uncollected garbage and overflowing sewers. Trash piles and pools of fetid water line the edges of the road. Iraqis stream through the smoky haze hauling random looted goods - ceiling fans, pieces of machinery, fluorescent lights, mismatched filing-cabinet drawers.

The bedlam continues until First Recon moves north of the city and links up with a light-armored reconnaissance company that is joining in the assault on Ba'qubah. The call sign of this adjoining company, which consists of about a hundred Marines mounted in twenty-four light-armored vehicles, is 'War Pig.' LAVs are noisy, black-armored eight-wheel vehicles shaped like upside-down bathtubs with rapid-fire cannons mounted on top. Iraqis call them "the Great Destroyers."

Despite the fact that Colbert's team has been driving into ambushes on an almost daily basis for more than two weeks, this is the first time these Marines have started a mission with an armored escort. "Damn! That's fucking awesome," Person says. "We've got the Great Destroyers with us."

"No, the escort is not awesome," Colbert says. "This just tells us how bad they're expecting this to be." As we pull out, Colbert's mood shifts from darkly brooding to grimly cheerful. "Once more into the great good night," he says in a mock stage voice, then quotes a line from Julius Caesar. "Cry 'havoc,' and let slip the dogs of war."

"Enemy contact," Colbert says, passing on word from his headset radio. "LAVs report enemy contact ahead."

War Pig is spread out on the highway, with its closest vehicle about a hundred meters directly in front of Colbert's and its farthest about three kilometers ahead. Automatic cannons send out tracer rounds that look like orange ropes. They stream out in all directions, orange lines bouncing and quivering over the landscape. Other, thinner orange lines, representing enemy machine guns, stream in toward the LAVs.

Iraqi Republican Guard troops have dug into trenches along both sides of the road. The enemy fighters are armed with every conceivable type of portable weapon - from machine guns to mortars to rocket-propelled grenades. The convoy stops as War Pig and the Iraqis shoot it out ahead. Enemy mortars explode nearby, falling from the sky in a random pattern. The Recon company behind Colbert's platoon opens up with everything it has. These Marines belong to a reservist unit, just arrived in Baghdad and only linked up with First Recon a few days earlier. They're older - a lot of them are beat cops or Drug Enforcement Administration agents in civilian life. This is their first significant enemy contact, and their wild firing - some of it in the direction of Colbert's Humvee - seems panicked.

"I have no targets! I have no targets!" Colbert repeats over the gunfire, but Corporal Walt Hasser, the gunner in the turret who operates the Mark-19 grenade launcher, begins lobbing rounds toward a nearby village.

"Cease fire!" Colbert shouts. "Easy there, buddy. You're shooting a village. We've got women and children there."

The reservists behind us have already poured at least a hundred grenades onto the small clusters of houses by the side of the road. In the window of one dwelling, a lantern glows. Through his night-vision scope, Colbert can just make out a group of what appears to be women and children taking cover behind a wall.

"We're not shooting the village, okay?" he says. In times like this, Colbert often assumes the tone of a schoolteacher calling a timeout during a frenzied playground scuffle. Mortars are exploding so close you feel the overpressure punching down on the Humvee. But Colbert will not allow his team to give in to the frenzy and shoot unless it finds clear targets or enemy muzzle flashes.

For the next twenty sleepless hours, the Marines in First Recon and War Pig methodically advance up the highway, traveling barely fifteen kilometers, clearing villages on foot, blowing up enemy trucks and weapons caches, and wiping out pockets of Iraqi soldiers as they hide in trenches or take cover in civilian homes.

From a raw-fear standpoint, the worst moments of the fight come early on the afternoon of April 9th. The world's attention is focused on televised pictures of American Marines in the center of Baghdad, pulling down a massive statue of Saddam Hussein. Here, north of the city, enemy mortars start exploding about thirty meters away from Bravo Company's position.

When Lieutenant Fick reports the bombardment to his commander over the radio, he is told to remain in position. "Stand by to die, gents," says Sergeant Antonio Espera, a former Los Angeles repo man and co-leader of the Humvee team that works in closest proximity to Colbert's. The twenty-two Marines in the platoon sit in their vehicles, engines running as per their orders, while mortars explode all around. There's almost no conversation. Everyone watches the sky and surrounding fields for mortar blasts. One lands five meters from Sergeant Espera's open-top Humvee, blowing a four-foot-wide hole in the ground.

I look out and see Espera hunched over his weapon, his eyes darting beneath the brim of his helmet, watching for the next hit. Beside him his driver, Corporal Jason Lilley, grips the wheel, his face ashen. A few hours before leaving on this mission Lilley had been sitting around with the platoon talking about the time he ate a clown fish - just for the hell of it - when he worked at a Wal-Mart in high school. Lilley joined the Marines to get out of his hometown in Wichita, Kansas, and stop partying. "My brains were, like, pan-fried," he says.

Before leaving on this mission, many of the men in Colbert's platoon had said goodbye to one another by shaking hands or even by hugging. The formal farewells seemed odd considering that everyone was going to be shoulder-to-shoulder in the cramped Humvees. The goodbyes almost seemed an acknowledgment of the transformations that take place in combat. Friends who lolled around together during free time talking about bands, girlfriends and eating clown fish aren't really the same people anymore once they enter the battlefield.

In combat, the change seems physical at first. Adrenaline begins to flood your system the moment the first bullet is fired. But unlike adrenaline rushes in the civilian world - a car accident or bungee jump, where the surge lasts only a few minutes - in combat, the rush can go on for hours. In time your body seems to burn out from it, or maybe the adrenaline just runs out. Whatever the case, after a while you begin to almost lose the physical capacity for fear. Explosions go off. You cease to jump or flinch. In this moment now, everyone sits still, numbly watching the mortars thump down nearby. The only things moving are the pupils of their eyes.

This is not to say the terror goes away. It simply moves out from the twitching muscles and nerves in your body and takes up residence in your mind. If you feed it with morbid thoughts of all the terrible ways you could be maimed or die, it gets worse. It also gets worse if you think about pleasant things. Good memories or plans for the future just remind you how much you don't want to die or get hurt. It's best to shut down, to block everything out. But to reach that state, you have to almost give up being yourself. This is why, I believe, everyone had said goodbye to each other. They would still be together, but they wouldn't really be seeing one another for a while, since each man would in his own way be sort of gone.

After about twenty minutes the mortar fire ceases for the rest of the day. Enemy resistance is beginning to wither under the combined effects of the Marine advance on the ground and violent air strikes from above. Had the Iraqis massed their armor earlier in the day when heavy clouds inhibited air strikes, they could have wreaked havoc. But for some reason, they missed their chance. Clouds have burned off, and waves of jets and Cobra helicopters simultaneously bomb, rocket and strafe targets in all directions. Trucks, armor, homes and entire hamlets are being bombed and set on fire. With the dramatic increase in firepower from the air, First Recon and War Pig rampage north, covering the final ten kilometers to Ba'qubah in a couple of hours. When the Iraqis finally send down a few armored vehicles, they are blown to smithereens by attack jets and Marines with shoulder-fired missiles.

The Iraqis who had put up fierce resistance earlier have either fled or been slaughtered. Headless corpses - indicating well-aimed shots from high-caliber weapons - are sprawled out in trenches by the road. Others are charred beyond recognition behind the wheels of burnt, skeletonized trucks. The sole injury on the American side occurs when a Marine in Alpha Company is hit by a piece of flying shrapnel from a

T-72 tank after it's blown up by one of his buddies with a shoulder-fired missile. His helmet, though partially crushed, stops the shrapnel. All the Marine suffered was a bad headache.

With each air assault, Recon teams advance into the flames and smoke, hunting for fleeing enemy fighters. The only people Colbert's team encounters are terrified villagers - a half-dozen men and one small, extremely frightened girl hiding in a ditch while their homes, fields and grape arbors burn in the wake of a Cobra attack. The men, fearing for their lives, scream, "No Saddam! No Saddam!" when Colbert's team approaches, weapons drawn. After Colbert and Fick pat the men on their shoulders to reassure them that they are not going to be executed, the village elder bursts into tears, grabs Fick's face and smothers him in kisses.

While this is going on, Sergeant Eric Kocher, leading a team in Bravo's Third Platoon on a sweep of a nearby field, bumps up against another group of Marines from the reserve Recon unit. About six of the reservists surround a dead enemy fighter, a young man in a ditch, lying in a pool of his own gore, still clutching his AK. While they ponder the corpse, Kocher apparently is the only one alert enough to notice a live Iraqi - this one armed - hiding in a trench nearby.

When Kocher alerts the reservist Marines to the presence of a live Iraqi in their midst, everyone turns his weapon on the man and shouts at him to stand up and drop his weapon. Ever since the weeklong battle in An Nasiriyah, where Iraqis attacked and killed Marines by luring them into ambushes with false surrenders, enemy takedowns have become highly charged affairs. One of the reservist Marines at the scene, First Sergeant Robert Cottle, a thirty-seven-year-old SWAT team instructor with the Los Angeles Police Department, takes out a pair of zip cuffs - sort of like heavy-duty versions of the plastic bands used to tie trash bags - and binds the Iraqi's hands behind his back.

Cottle cuffs the enemy prisoner's wrists so tightly that his arms later develop dark-purple blood streaks all the way to his shoulders. The prisoner, a low-level Republican Guard volunteer in his late forties, is overweight, dressed in civilian clothes - a sleeveless undershirt and filthy trousers - and has a droopy Saddam mustache. He looks like a guy so out of shape, he'd get winded driving a taxicab in rush hour. Surrounded by Marines, the man begins to blubber and cry.

A few minutes later Cottle, the reservist, shook Kocher's hand and thanked him for spotting the Iraqi and said, "You might have just saved our lives."

EXTREME MARINES

By Michael W. Rodriguez

"The feminization of the military has been, and remains, a cancer eating away at the warrior spirit and preparedness of the United States military." – Geoff Metcalf

Secretary of Defense Donald Rumsfeld took a lot of heat for some of his comments, but I for one found his straightforward, no-nonsense approach refreshing. It wasn't long ago his office was occupied by a bunch of double-talking bureaucrats with a penchant for reading Karl Marx at bedtime, and those just aren't the sort of people I want guarding our borders.

Of all the laughable, left-wing Clinton appointees, one stood out in the crowd as being particularly heinous. Sara Lister had about as much business serving in the Department of Defense as I would have designing ladies evening wear, and her appointment made about as much sense as making a ballerina the linebackers coach for the Green Bay Packers. I wish I could have been there when General Krulak called her to task:

Sara E. Lister, the former Assistant Secretary of the Army and an ardent advocate of women in combat, called Marines "extremists" and "a little dangerous" in a public statement back in 1997.

"I think the Army is much more connected to society than the Marines are," Ms. Lister told a seminar on October 26, 1997. "The Marines are extremists. Wherever you have extremists, you've got some risks of total disconnection with society. And that's a little dangerous."

Commandant of the Marine Corps, General Charles Krulak, fired back. "Assistant Secretary of the Army Sara Lister has been quoted as characterizing the Marine Corps as 'extremists,'" General Krulak said. "Such a depiction would summarily dismiss 222 years of sacrifice and dedication to the nation. It would dishonor the hundreds of thousands of Marines whose blood has been shed in the name of freedom. Citizens from all walks of life have donned the Marine Corps uniform, and gone to war to defend this nation, never to return. Honor, courage and commitment are not extreme."

I have to assume that Ms. Lister includes all Marines who have ever worn the, quote, "checkerboard fancy uniforms and stuff..." that she mocked in her speech. If she does, and clearly that was her intent, I have to address her remarks, and then make a few of my own.

Are Marines 'extremists'? Dr. Daniel W. Polland, a former Navy Corpsman in the 1st Battalion, 9th Marines and now a physician in Colorado said this at a gathering of veterans:

"Oh Lord, we have long known that prayer should include confession. Therefore on behalf of the Marines... I confess their sins.

Lord, they're just not in step with today's society. They are unreasonable in clinging to old-fashioned ideas like patriotism, duty, honor, and country. They hold radical ideas, believing that they are their brother's keeper and responsible for the Marine on their flank.

They have been seen standing when colors pass, singing the National Anthem at ball games, and drinking toasts to fallen comrades. Not only that, they have frequently been observed standing tall, taking charge and wearing their hair unfashionably short.

They have taken John Kennedy's words too seriously and are overly concerned with what they can do for their country, instead of what this country can do for them. They take the Pledge of Allegiance to heart and believe that their oath is to be honored.

Forgive them, Lord, for being stubborn men and women who hold fast to such old-fashioned values. After all, what more can you expect? They're Marines!

And a former Air Force officer, Vietnam veteran and fighter pilot named Toby Hughes added in response to Ms. Lister's ill-thought remarks, "…I never knew many Marines who were only a 'little' dangerous. Most of them seem to be a LOT dangerous. That, I think, is the idea."

Are Marines as 'extremist' as Ms. Lister would have us believe? Yes, Ms. Lister, we are. We are extreme in our pride. We are extreme, even insufferable, in our Brotherhood. We are extreme in our devotion to duty, to service, to country, to you.

After all, the Chief of Naval Operations would never be called a Sailor. The General of the Army would never be called a Soldier. The Chief of Staff of the Air Force would never be called an Airman. But the Commandant of the Marine Corps is damned proud to be called Marine.

We will always be 'extreme.' We know no other way to live, to serve, to pay back our country for what we have and for what we have earned. And what we have earned, Ms. Lister, is a title. A name. A right. What we have earned is, I believe, something which you will never understand. What we have earned you can never take from us.

DEAD TO RIGHTS

"Few things can help an individual more than to place responsibility on him, and to let him know you trust him." - Booker T. Washington

We have all heard the phrase, "good initiative, poor judgment," but I learned about its meaning the hard way. When I was a young corporal I screwed up while doing what I thought was right, and that mistake can now serve as an example to others - and as proof that you can get into trouble and still have a successful career in the Marine Corps.

Back it the seventies it was still a big deal to become a non-commissioned officer, so when I got promoted to corporal I took my responsibilities very seriously. That became even more true after I attended NCO Leadership School - I was a hard charger! More than anything else I took to heart the need for an NCO to watch out for his men's welfare, to mold them into a productive part of the team, and to take a real interest in the lives of each man in his charge.

And so it was that I found myself pulling a tour as the Duty NCO on a Sunday night for Company A, 2nd Reconnaissance Battalion out at Onslow Beach in the far reaches of Camp Lejeune. Anyone who ever served there remembers those old barracks - large open squad bays in buildings which were rumored to have housed POWs during WWII. In a word, they were… luxurious!

That particular weekend my platoon was tasked with acting as aggressors against one of the other companies, and someone in the brain trust decided it would be better for us to store our M-16s in our wall lockers for the weekend rather than keep the armory open. Big mistake.

Late that evening I learned that one of my platoon-mates, a private named ‘Moon,’ had become extremely intoxicated and had passed out at the far end of the squad bay. I wasn’t able to wake him up at first, and decided it would be a good idea to toss him in the shower in order to bring him around. It worked, but before long I would end up regretting waking him.

I escorted Moon back to his rack and I told him to turn in for the night, which he did. For the next hour or so everything was peaceful and calm, until all of a sudden I heard a loud commotion coming from the vicinity of Moon’s rack. When I arrived I saw that he had pulled his weapon from his locker, and when I came into view he immediately pointed it in my direction. I wasn’t too concerned though, since all we had been issued that weekend were blanks. That’s when Moon pulled out a magazine which clearly contained at least one round of live ammo and proceeded to lock and load. NOW I was concerned!

At this point I began talking to the inebriated Marine in an attempt to calm him down. I knew that his father was a senior NCO in the Air Force, and partly for that reason I had taken him on as something of a reclamation project and had spent a lot of time over the previous few weeks trying to square him away. But he didn’t want to listen, and things began to go from bad to worse - and the entire time the weapon never wavered from the center of my chest. By now my Assistant Duty had gotten wind of what was going on, and he slowly worked his way around behind the lockers and came up behind the Private Moon - and when the moment was right swooped in and snatched the rifle away from him. Whew!

The following morning I pulled my platoon sergeant aside and filled him in on what had happened so he could deal with Moon, who by now had a pretty bad hangover. Imagine my surprise when I was the one brought up on charges - for dereliction of duty and disobeying a written order (for not making a logbook entry about the incident). Believe it or not I was even reprimanded for NOT beating the drunken private senseless once he had been disarmed.

I eventually was brought before the Company Commander for Non-Judicial Punishment, or what was more commonly known as Office Hours. I realized I had to take my lumps and vowed to learn from the whole affair, but to this day I can't help but be amazed at my CO's logic. He reasoned that since I was his direct representative in my capacity as Duty NCO Moon had actually been pointing that rifle at HIM. Now I can certainly understand that concept in an abstract sort of way, but the fact remained I would have been the one filling a body bag if Moon had decided to pull the trigger. Fortunately that didn't turn out to be the case.

The entire incident, while unfortunate, did teach me a lot, and I put those lessons to good use many times over the course of the next twenty years. It's ironic, but getting Office Hours made me a much better leader than I would have been otherwise - although I certainly don't recommend going that route as a means of self-improvement!

JOE FOSS

"The story of Joe Foss's life is a story of human endeavor so great and so accomplished that it defies exaggeration." - Senator John McCain

I wonder how many people looked at the title of this story and said "Who?" Most people of my generation don't know who Joe Foss was, and I would be shocked to find anyone under the age of thirty with a clue as to his identity. Ask young people today to name a war hero and they may very well say Jessica Lynch. How sad. Joe Foss was much more than just a war hero, and I was reminded of that when I learned of his recent death. He was an AMERICAN in every sense of the word. He once again came into the public eye not long before his death when he was subjected to an extensive search at an airport security checkpoint while on his way to address the Corps of Cadets at West Point. Metal detectors alerted the screeners to an object in Foss' pocket, which turned out to be his Medal of Honor. Not only did the security personnel not know who Foss was, or recognize the object in his pocket as our Nation's highest award for valor, they actually thought it might be some sort of weapon. Joe Foss, holder of the Medal of Honor and former Governor of South Dakota, a man who was in his eighties at the time, was detained as "suspicious" and made to remove his belt and boots several times. The point is he did not complain about not being recognized or about how he was treated personally (as many of today's "celebrities" would have), but instead was sad that the Medal of Honor was not recognized.

In tribute to his passing I think it is appropriate to tell his story:

Joe Foss was born in 1915 to a Norwegian-Scots family in South Dakota, where he learned hunting and marksmanship at a young age. Like millions of others, eleven-year-old Joe Foss was inspired by Charles Lindbergh, especially after he saw Lindy at an airport near Sioux Falls. Five years later he watched a Marine squadron put on a dazzling exhibition, led by Captain Clayton Jerome, future wartime Director of Marine Corps Aviation.

In 1934 Joe began his college education in Sioux Falls, but he had to drop out to help his mother run the family farm. However he somehow managed to scrape up sixty-five dollars for private flying lessons. Five years later he entered the University of South Dakota again, and supported himself by waiting on tables. In his senior year he also completed a civilian pilot training program before he graduated with a Business degree in 1940.

Upon graduation he enlisted in the Marine Corps reserves as an aviation cadet. Seven months later, he earned his Marine wings at Pensacola and was commissioned a second lieutenant. For the next nine months he was a 'plowback' flight instructor. He was at Pensacola when the news of Pearl Harbor broke, and since he was Officer of the Day was placed in charge of base security. Thus he prepared to defend Pensacola from Jap invaders, riding around the perimeter on a bicycle. To his distress he was then ordered to the aerial photographers school and assigned to VMO-1, a photo reconnaissance squadron, but he insisted he wanted fighter pilot duty even after being told "You're too ancient, Joe. You're twenty-seven years old!" After lengthy lobbying with Aircraft Carrier Training Group he learned all about the new

F4F Wildcat, logging over 150 flight hours in June and July, and when he finished training he became the executive officer of VMF-121. Three weeks later he was on his way to the South Pacific, where Americans were desperately trying to turn the tide of war, and upon arrival VMF-121 was loaded aboard the escort carrier *Copahee*.

On the morning of October 9 they were catapulted off the decks, in Joe's only combat carrier mission. Landing at Henderson Field, he was told that his fighters were now based at the 'cow pasture.' He was impressed with the make-do character of the 'Cactus Air Force.' The airfield was riddled with bomb craters and wrecked aircraft, but also featured three batteries of 90mm anti-aircraft guns and two radar stations. As exec of VMF-121 he would normally lead a flight of two four-plane divisions, whenever there were enough Wildcats to go around. He was the oldest pilot in the flight, four years older than the average age of twenty-three. The flight would become known as 'Foss's Flying Circus' and rack up over sixty victories. Five of them would become aces, and two would die in the in the fight for Guadalcanal.

On October 13, 1942 VMF-121 scored its first victories when Lieutenants Freeman and Narr each got a Japanese plane. Later that same day, Foss led a dozen Wildcats to intercept thirty-two enemy bombers and fighters. In his first combat a Zero bounced Joe, but overshot, and Joe was able to fire a good burst and claim one destroyed aircraft. Instantly three more Zeros set upon him, and he barely made it back to 'Fighter One,' with his Wildcat dripping oil. Chastened by the experience, he declared "You can call me 'Swivel-Neck Joe' from now on." From the first day, Joe followed the tactics of Joe Bauer: getting in close, so close that another pilot joked that the 'exec' left powder burns on his targets. The next day while intercepting a flight of enemy bombers Joe's engine acted up, and he took cover in the clouds. But suddenly a Wildcat whizzed past him, tailed by a Zero. Joe cut loose and shot the Zero's wing off. It was his second victory in two days.

While the Wildcats' primary responsibility was air defense, they also strafed Japanese infantry and ships when they had enough ammunition. Joe led one such mission on the 16th. Mid-October was the low point for the Americans in the struggle for Guadalcanal. Japanese warships shelled U.S. positions nightly, with special attention to the airstrips. To avoid the shelling, some fliers slept on the front lines. Foss grew to appreciate the Navy's fighter doctrine and found that the "Thach Weave" effectively countered the Zero's superior performance, because "it allowed us to point eyes and guns in every direction."

Joe was leading an interception on morning of the 18th when the Zero top cover pounced on them and downed an F4F. But Foss was able to get above them and flamed the nearest, hit another, and briefly engaged a third. Gaining an angle, he finally shot up the third plane's engine. Next he found a group of Bettys already under attack by VF-71. He executed a firing pass from above, flashed through the enemy bombers, and pulled up sharply, blasting one from below. Nine days at Guadalcanal and he was an ace! Two days later Lieutenant Colonel Bauer and Foss led a flight of Wildcats on the morning intercept. In the dogfighting, Joe downed two Zeros, but took a hit in his engine. He landed safely at Henderson Field with a bad cut on his head, but was otherwise unharmed.

'Cactus Fighter Command' struggled to keep enough Wildcats airworthy to meet the daily Japanese air strikes. On the 23rd it put up two flights, led by Foss and Major Davis. There were plenty of targets, and Joe soon exploded a Zero.

He went after another which tried to twist away in a looping maneuver. Joe followed and opened up while inverted at the top of his loop. He caught the Zero and flamed it. He later described it as a lucky shot. Next he spotted a Japanese pilot doing a slow roll; he fired as the Zero's wings rolled through the vertical and saw the enemy pilot blown out of the cockpit, minus a parachute. Suddenly he was all alone and two Zeros hit him, but his rugged Grumman absorbed the damage, permitting Foss to flame one of his assailants. Once again, he nursed a damaged fighter back to Guadalalcanal. So far he had destroyed eleven enemy planes, but had brought back four Wildcats that were too damaged to fly again.

October 25 was the day the Japanese planned to occupy Henderson Field; they sent their fighters over with orders to circle until the airstrip was theirs. It didn't work out that way, as the U.S. ground forces held their lines and 'Cactus did its part. Joe Foss led six Wildcats up before 10 AM, and claimed two of the Marines' three kills on that sortie. In an afternoon mission on the 25th, he downed three more, to become the Marine Corps' first 'ace in a day.' He had achieved fourteen victories in only thirteen days.

On November 7th Foss led seven F4Fs up the 'Slot' to attack some destroyers and a cruiser which were covered by six Rufe floatplane fighters. They dispatched five of the Rufes promptly and prepared to strafe the destroyers. Joe climbed up to protect the others and got involved in a dogfight with a Pete, a two-man float biplane. He shot down the slow-flying plane, but not before its rear gunner perforated the Wildcat's engine with 7.7mm machine gun fire. Once again Foss' aircraft started sputtering on the way home, but his time it didn't make it. As the engine died he put it into the longest possible shallow dive, to get as close to home as he could.

As the plane went into the water off Malaita Island, Foss struggled with his parachute harness and his seat. He went under with his plane, gulped salt water, and almost drowned before he freed himself and inflated his Mae West. Exhausted and with the tide against him, he knew that he couldn't swim to shore. While trying to rest and re-gain his strength in his life raft, he spotted shark fins nearby. He sprinkled the chlorine powder supplied for that purpose in his emergency pack and that seemed to help. As darkness approached, he heard some searchers looking for him. They hauled him in and brought him to Malaita's Catholic mission. There were a number of Europeans and Australians there, including two nuns who had been there for forty years and had never seen an automobile. They fed him steak and eggs and invited him stay for two weeks.

The next day a PBY Catalina rescued him, and on his return to Guadalcanal he learned that 'Cactus' had downed fifteen Japanese planes in the previous day's air battle. His own tally stood at nineteen, and on the 9th of November Admiral Bull Halsey pinned the Distinguished Flying Cross on Foss and two other pilots.

The Americans were bringing four transports full of infantry to Guadalcanal three days later, and the Japanese sent sixteen Betty bombers and thirty covering Zeroes after them while the American Wildcats and Airacobras defended. Foss and his Wildcats were flying top cover CAP and dove headlong into the attackers, right down onto the deck. As Barrett Tillman described it in *Wildcat Aces of WWII:* "Ignoring the peril, Foss hauled into within one hundred yards of the nearest bomber and aimed at the starboard engine, which spouted flame. he G4M tried a water landing, caught a wingtip and tumbled into oblivion. Foss set his sight on another Betty when a Zero intervened. The F4F nosed up briefly and fired a beautifully aimed snapshot which sent the A6M spearing into the water. He then resumed the chase."

Foss caught up with the next Betty in line and made a deflection shot into its wingroot; the bomber flamed up and then set down in the water. The massive dogfight continued, until Joe ran out of fuel and ammunition.

Late that afternoon Colonel Bauer, tired of being stuck on the ground at Fighter Command, went up with Joe to take a look. It was his last flight, described by Joe Foss in a letter to Bauer's family. No trace of 'Indian Joe' was ever found. Back at Guadalcanal, Foss was diagnosed with malaria. The two great leaders of Cactus Fighter Command were gone, although Foss would return in six weeks.

Foss returned to combat flying on the 15th when he shot down three more planes to bring his total to twenty-six. He flew his last mission ten days later when his flight and four P-38s intercepted a force of over sixty Zeros and Vals. Quickly analyzing the situation, he ordered his flight to stay high, circling in a Lufbery. This made his small flight look like a decoy to the Japanese. Soon Cactus scrambled more fighters and the Japanese planes fled. It was ironic that in one of Joe Foss' most satisfying missions, he didn't fire a shot.

A few months later he went to Washington D.C., to be decorated and begin "the dancing bear act" for his twenty-six aerial victories – which had equaled Eddie Rickenbacker's World War One record. He gave pep talks, made factory tours, and went on the inevitable War Bond tours. In May of 1943 President Roosevelt presented him with the Medal of Honor for outstanding heroism above and beyond the call of duty.

After the war Foss was commissioned in the South Dakota Air National Guard, which he helped organize. Joe then turned to politics and was elected to the South Dakota House of Representatives, and during the Korean War he returned to active duty as an Air Force Colonel. Foss later became the chief of staff of the South Dakota Air National Guard with the rank of Brigadier General, and in 1954 he was overwhelmingly elected Governor of South Dakota - and two years later was elected to a second term. After that he was elected the first commissioner of the American Football League, and served until 1966. He was also president of the National Rifle Association (NRA) from 1988 - 1990, and was featured in Tom Brokaw's best-seller *The Greatest Generation.* Joe Foss was *much* more than "just" a fighter pilot!

The Joe Foss Institute is on the internet at www.thefossinstitute.org

LOATHING THE MILITARY

LtCol Michael Mark

"A man who will not protect his freedom does not deserve to be free." - General Douglas MacArthur

As the 2004 elections grew near it was appropriate for those of us who have served in the Armed Forces to take a look back at William Jefferson Clinton's tenure not as President, but as Commander-in-Chief. That such a man served in that capacity is one of the great ironies in history. And as the specter of Hillary the Presidential candidate began to loom we needed to ask ourselves if she was more, or less, qualified to lead our nation's military than her husband. There is no doubt she was sympathetic to his actions, but that is a story for another day.

The Bill Clinton story is one of cowardice and deceit... of a twenty-three-year-old man, not some terrified and confused eighteen-year-old boy, manipulating the system to avoid his duty. One can read the level of his smug contempt for the military and those who served in his letter. We will never know which mother's son took his place on the battlefield and if he ever came home. Incidentally, Clinton was never graduated from Oxford, yet he allows himself to be referred to as a Rhodes scholar, giving the inference he'd graduated - just another glimpse at his lack of integrity. It was party time and anti-war demonstrations for Bill while at Oxford; then on to Moscow, a subject that remains a dark secret and may just involve his wartime service... but for which side?

When his undergraduate draft deferment ran out and he went off to Oxford he sought further protection from the draft by promising the ROTC Commandant at the University of Arkansas that he'd return in the fall of 1969 and enroll in ROTC.

He had already received and ignored a draft notice. An ROTC deferment was granted, yet it shouldn't have been because he wasn't enrolled, but Bill had connections with the powerful Senator Fulbright of Arkansas. He mailed this letter to Colonel Holmes, ROTC Commandant, after he was awarded a draft number that assured he wouldn't be called to serve.

Following is the text of the letter that Bill Clinton wrote to Colonel Eugene Holmes, director to the ROTC program at the University of Arkansas, on December 3rd, 1969:

Dear Colonel Holmes,

I'm sorry to be so long in writing. I know I promised to let you hear from me at least once a month, and from now on you will, but I have had some time to think about this first letter. Almost daily since my return to England I have thought about writing, about what I want to and ought to say. First, I want to thank you, not just for saving me from the draft, but for being so kind and decent to me last summer, when I was as low as I have ever been. One thing which made the bond we struck in good faith somewhat palatable to me was my high regard for you personally. In retrospect it seems that the admiration might not have been mutual had you known a little more about me, about my political beliefs and activities. At least you might have thought me more fit for the draft than for ROTC.

Let me try to explain. As you know I worked for two years in a very minor position on the State Foreign Relations Committee. I did it for the experience and the salary but also for the opportunity, however small, of working every day against a war I opposed and despised with a depth of feeling I had reserved solely for racism in America before Vietnam. I did not take the matter lightly but studied it carefully, and there was a time when not many people had more information about Vietnam at hand than I did.

I have written and spoken and marched against the war. One of the national organizers of the Vietnam Moratorium is a close friend of mine. After I left Arkansas last summer I went to Washington to work in the national headquarters of the Moratorium, then to England to organize the Americans here for demonstrations 16 Oct and 16 Nov.

Interlocked with the war is the draft issue, which I did not begin to consider separately until early 1968. After a law seminar at Georgetown I wrote a paper on the legal arguments for and against allowing, within the Selective Service System, the classification of selective conscientious objection, for those opposed to participation in a particular war, not simply to "participation in war in any form."

From my work I came to believe that the draft system itself is illegitimate. No government really rooted in limited parliamentary democracy should have the power to make its citizens fight and kill and die in a war they may oppose, a war which even possibly may be wrong, a war which, in any case, does not involve immediately the peace and freedom of the nation. The draft was justified in World War II because the life of the people collectively was at stake. Individuals had to fight, if the nation was to survive, for the lives of their countrymen and their way of life. Vietnam is no such case. Nor was Korea an example where, in my opinion, certain military action was justified but the draft was not for the reasons stated above.

Because of my opposition to the draft and the war, I am in great sympathy with those who are not willing to fight, kill, and maybe, die for their country (i.e. the particular policy of a particular government) right or wrong. Two of my friends at Oxford are conscientious objectors. I wrote a letter of recommendation for one of them to his Mississippi draft board, a letter which I am more proud of than anything else I wrote at Oxford last year. One of my roommates is a draft resister who is possibly under indictment and may never be able to go home again. He is one of the bravest, best men I know. His country needs men like him more than they know. That he is considered a criminal is an obscenity.

The decision not to be a resister and the related subsequent decisions were the most difficult of my life. I decided to accept the draft in spite of my beliefs for one reason: to maintain my political viability within the system. For years I have worked to prepare myself for a political life characterized by both practical political ability and concern for rapid social progress. It is a life I still feel compelled to try to lead. I do not think our system of government is by definition corrupt, however dangerous and inadequate it has been in recent years. (The society may be corrupt, but that is not the same thing, and if that is true we are all finished anyway.)

When the draft came, despite political convictions, I was having a hard time facing the prospect of fighting a war I had been fighting against, and that is why I contacted you. ROTC was the one way left in which I'd possibly, but not positively, avoid both Vietnam and resistance. Going on with my education, even coming back to England, played no part in my decision to join ROTC. I am back here and would have been at Arkansas Law School because there is nothing else I can do. In fact, I would like to have been able to take a year out perhaps to teach in a small college or work on some community action project and in the process to decide whether to attend law school or graduate school and how to begin putting what I have learned to use.

But the particulars of my personal life are not nearly as important to me as the principles involved. After I signed the ROTC letter of intent, I begin to wonder whether the compromise I had made with myself was not more objectionable than the draft would have been, because I had no interest in the ROTC program in itself and all I seemed to have done was to protect myself from physical harm. Also, I began to think I had deceived you, not by lies - there were none - but by failing to tell you all the things I'm writing now. I doubt that I had the mental coherence to articulate them then.

At that time, after we had made our agreement and you had sent my 1-D deferment to my draft board, the anguish and loss of my self-regard and self-confidence really set in. I hardly slept for weeks, and kept going by eating compulsively and reading until exhaustion brought sleep. Finally, on 12 September, I stayed up all night writing a letter to the chairman of my draft board, saying basically what is in the preceding paragraph, thanking him for trying to help in a case where he really couldn't, and stating that I couldn't do the ROTC after all and would he please draft me as soon as possible. I never mailed the letter, but I did carry it on me every day until I got on the plane to return to England. I didn't mail the letter because I didn't see, in the end, how my going in the army and maybe to Vietnam would achieve anything except a feeling that I had punished myself and gotten what I deserved. So I came back to

England to try to make something of this second year of my Rhodes scholarship. And that is where I am now, writing to you because you have been good to me and have a right to know what I think and feel. I am writing too, in the hope my telling this one story will help you to understand more clearly how so many fine people have come to find themselves still loving their country but loathing the military, to which you and other good men have devoted years, lifetimes, of the best service you could give. To many of us, it is no longer clear what is service and what is disservice, or if it is clear, the conclusion is likely to be illegal.

Forgive the length of this letter. There was much-to say. There is still a lot to be said, but it can wait. Please say hello to Colonel Jones for me. Merry Christmas.

Sincerely,
Bill Clinton

This article originally appeared in Military Magazine

SOME THOUGHTS
On Personal Responsibility

"The Constitution only gives people the right to *pursue* happiness. You have to catch it yourself."- Benjamin Franklin

These days it is almost impossible to get off a highway exit without encountering a panhandler at the stoplight holding a sign proclaiming himself to be a "Homeless Vet." For a long time those signs said "Vietnam Vet," but as the bums in this country get younger it became necessary for them to make a change. Either way, I am offended by the implication that veterans are unable to care for themselves. No doubt there are some former servicemen among the ranks of the homeless, but for the most part these people are human refuse who have never served their country - and who use the average person's respect for the military to finance their drinking binges. I have called their bluff on a number of occasions by offering a job in lieu of money, but have had no takers as of yet. A new tactic in recent years has been for them to post a sign that says "I won't lie, it's for booze," with the expectation of being rewarded for their honesty. Any way you look at it, these people are the product of a society in which people avoid taking responsibility for their own lives, and blame their plight on others. Those people tend to refer to themselves as "liberals."

Liberals love to portray the Republican Party as the "party of the rich," which makes some sense on the surface. People with a strong work ethic tend to become successful, and eventually some of them become wealthy, and conservative. In the meantime the ranks of the Democratic

Party are filled with liberal college professors and those on welfare. Those demographics are of course not without exception, but you get the idea.

The thing that blows the whole "party of the rich" theory out of the water for me is the fact that our military members overwhelmingly favor the conservative Right, and anyone who has ever drawn a paycheck from Uncle Sam knows there are no fortunes to be made in the Armed Forces. But perhaps the liberals who spawned that theory inadvertently stumbled upon the truth - conservatives ARE "rich," but not necessarily in the material sense. Instead they are rich in values, in dedication, in morals, and in personal responsibility.

A nationally syndicated talk show host from Atlanta named Neal Boortz - who happens to be the son of a Marine - recently penned a "commencement address" on the subject, and whether you agree or not you will find his views thought provoking. I think his comments should be taken to heart by everyone from the Commandant down to the newest recruit. – and it would have been particularly entertaining to witness the faculty's reaction when he spoke:

"I am honored by the invitation to address you on this august occasion. It's about time. Be warned, however, that I am not here to impress you; you'll have enough smoke blown your way today. And you can bet your tassels I'm not here to impress the faculty and administration. You may not like much of what I have to say, and that's fine. You will remember it though. Especially after about ten years out there in the real world. This, it goes without saying, does not apply to those of you who will seek your careers and your fortunes as government employees.

This gowned gaggle behind me is your faculty. You've heard the old saying that those who can - do. Those who can't - teach. That sounds deliciously insensitive. But there is often raw truth in insensitivity, just as you often find feel-good falsehoods and lies in compassion. Say good-bye to your faculty because now you are getting ready to go out there and do. These folks behind me are going to stay right here and teach.

By the way, just because you are leaving this place with a diploma doesn't mean the learning is over. When an FAA flight examiner handed me my private pilot's license many years ago, he said, 'Here, this is your ticket to learn.' The same can be said for your diploma. Believe me, the learning has just begun.

Now, I realize that most of you consider yourselves Liberals. In fact, you are probably very proud of your liberal views. You care so much. You feel so much. You want to help so much. After all, you're a compassionate and caring person, aren't you now? Well, isn't that just so extraordinarily special! Now, at this age, is as good a time as any to be a Liberal; as good a time as any to know absolutely everything. You have plenty of time, starting tomorrow, for the truth to set in. Over the next few years, as you begin to feel the cold breath of reality down your neck, things are going to start changing pretty fast - including your own assessment of just how much you really know.

So here are the first assignments for your initial class in reality: Pay attention to the news, read newspapers, and listen to the words and phrases that proud Liberals use to promote their causes. Then compare the words of the Left to the words and phrases you hear from those evil, heartless, greedy conservatives. From the Left you will hear "I feel." From the Right you will hear "I think." From the Liberals you will hear references to groups - The Blacks, The Poor, The Rich, The Disadvantaged, The Less Fortunate. From the Right you will hear references to individuals. On the Left you hear talk of group rights; on the Right, individual rights. That about sums it up, really: Liberals feel. Liberals care. They are pack animals whose identity is tied up in group dynamics. Conservatives and Libertarians think - and, setting aside the theocracy crowd, their identity is centered on the individual.

Liberals feel that their favored groups have enforceable rights to the property and services of productive individuals. Conservatives (and Libertarians, myself among them I might add) think that individuals have the right to protect their lives and their property from the plunder of the masses.

In college you developed a group mentality, but if you look closely at your diplomas you will see that they have your individual names on them. Not the name of your school mascot, or of your fraternity or sorority, but your name. Your group identity is going away. Your recognition and appreciation of your individual identity starts now.

If, by the time you reach the age of thirty, you do not consider yourself to be a libertarian or a conservative, rush right back here as quickly as you can and apply for a faculty position. These people will welcome you with open arms. They will welcome you, that is, so long as you haven't developed an individual identity. Once again you will have to be willing to sign on to the group mentality you embraced during the past four years.

Something ele is going to happen soon that is going to really open your eyes. You're going to actually get a full time job! You're also going to get a lifelong work partner. This partner isn't going to help you do your job. This partner is just going to sit back and wait for payday. This partner doesn't want to share in your effort, just your earnings.

Your new lifelong partner is actually an agent. An agent representing a strange and diverse group of people. An agent for every teenager with an illegitimate child. An agent for a research scientist who wants to make some cash answering the age-old question of why monkeys grind their teeth. An agent for some poor demented hippie who considers herself to be a meaningful and talented artist… but who just can't manage to sell any of her artwork on the open market.

Your new partner is an agent for every person with limited, if any, job skills… but who wanted a job at City Hall. An agent for tin-horn dictators in fancy military uniforms grasping for American foreign aid. An agent for multi-million-dollar companies who want someone else to pay for their overseas advertising. An agent for everybody who wants to use the unimaginable power of this agent for their personal enrichment and benefit.

That agent is our wonderful, caring, compassionate, oppressive government. Believe me, you will be awed by the unimaginable power this agent has. Power that you do not have. A power that no individual has, or will have. This agent has the legal power to use force deadly force to accomplish its goals.

You have no choice here. Your new friend is just going to walk up to you, introduce itself rather gruffly, hand you a few forms to fill out, and move right on in. Say hello to your own personal one ton gorilla. It will sleep anywhere it wants to.

Now, let me tell you, this agent is not cheap. As you become successful it will seize about forty percent of everything you earn. And no, I'm sorry, there just isn't any way you can fire this agent of plunder, and you can't decrease its share of your income. That power rests with him, not you.

So, here I am saying negative things to you about government. Well, be clear on this: It is not wrong to distrust government. It is not wrong to fear government. In certain cases it is not even wrong to despise government for government is inherently evil. Yes… a necessary evil, but dangerous nonetheless… somewhat like a drug. Just as a drug that in the proper dosage can save your life, an overdose of government can be fatal.

Now let's address a few things that have been crammed into your minds at this university. There are some ideas you need to expunge as soon as possible. These ideas may work well in academic environment, but they fail miserably out there in the real world.

First that favorite buzz word of the media, government, and academia: Diversity! You have been taught that the real value of any group of people - be it a social group, an employee group, a management group, whatever - is based on diversity. This is a favored liberal ideal because diversity is based not on an individual's abilities or character, but on a person's identity and status as a member of a group. Yes it's that liberal group identity thing again.

Within the great diversity movement group identification - be it racial, gender based, or some other minority status - means more than the individual's integrity, character or other qualifications.

Brace yourself. You are about to move from this academic atmosphere where diversity rules, to a workplace and a culture where individual achievement and excellence actually count. No matter what your professors have taught you over the last four years, you are about to learn that diversity is absolutely no replacement for excellence, ability, and individual hard work. From this day on every single time you hear the word "diversity" you can rest assured that there is someone close by who is determined to rob you of every vestige of individuality you possess.

We also need to address this thing you seem to have about "rights." We have witnessed an obscene explosion of so-called "rights" in the last few decades, usually emanating from college campuses.

You know the mantra: You have the right to a job. The right to a place to live. The right to a living wage. The right to health care. The right to an education. You probably even have your own pet right - the right to a Beemer, for instance, or the right to have someone else provide for that child you plan on downloading in a year or so.

Forget it. Forget those rights! I'll tell you what your rights are! You have a right to live free, and to the results of your labor. I'll also tell you have no right to any portion of the life or labor of another.

You may, for instance, think that you have a right to health care. After all, Hillary said so, didn't she? But you cannot receive health care unless some doctor or health practitioner surrenders some of his time - his life - to you. He may be willing to do this for compensation, but that's his choice. You have no "right" to his time or property. You have no right to his or any other person's life or to any portion thereof.

You may also think you have some "right" to a job; a job with a living wage, whatever that is. Do you mean to tell me that you have a right to force your services on another person, and then the right to demand that this person compensate you with their money? Sorry, forget it. I am sure you would scream if you started your own business and some urban outdoorsmen (that would be "homeless person" for those of you who don't want to give these less fortunate people a romantic and adventurous title) came to you and demanded his job and your money.

The people who have been telling you about all the rights you have are simply exercising one of theirs - the right to be imbeciles. Their being imbeciles didn't cost anyone else either property or time. It's their right, and they exercise it brilliantly.

By the way, did you catch my use of the phrase "less fortunate" a bit ago when I was talking about the urban outdoorsmen? That phrase is a favorite of the Left. Think about it, and you'll understand why.

To imply that one person is homeless, destitute, dirty, drunk, spaced out on drugs, unemployable, and generally miserable because he is "less fortunate" is to imply that a successful person - one with a job, a home and a future - is in that position because he or she was "fortunate." The dictionary says that fortunate means "having derived good from an unexpected place." There is nothing unexpected about deriving good from hard work. There is also nothing unexpected about deriving misery from choosing drugs, alcohol, and the street.

If the Left can create the common perception that success and failure are simple matters of "fortune" or "luck," then it is easy to promote and justify their various income redistribution schemes. After all, we are just evening out the odds a little bit.

This "success equals luck" idea the liberals like to push is seen everywhere. Democratic presidential candidate Richard Gephardt referred to high-achievers as "people who have won life's lottery." He wants you to believe they are making the big bucks because they are lucky.

It's not luck, my friends. It's choice. One of the greatest lessons I ever learned was in a book by Og Mandino, entitled *The Greatest Secret in the World.* The lesson? Very simple: "Use wisely your power of choice."

That bum sitting on a heating grate, smelling like a wharf rat? He's there by choice. He is there because of the sum total of the choices he has made in his life. This truism is absolutely the hardest thing for some people to accept, especially those who consider themselves to be victims of something or other - victims of discrimination, bad luck, the system, capitalism, whatever. After all, nobody really wants to accept the blame for his or her position in life. Not when it is so much easier to point and say, "Look! He did this to me!" than it is to look into a mirror and say, "You S.O.B.! You did this to me!"

The key to accepting responsibility for your life is to accept the fact that your choices, every one of them, are leading you inexorably to either success or failure, however you define those terms.

Some of the choices are obvious: Whether or not to stay in school. Whether or not to get pregnant. Whether or not to hit the bottle. Whether or not to keep this job you hate until you get another better-paying job. Whether or not to save some of your money, or saddle yourself with huge payments for that new car.

Some of the choices are seemingly insignificant: Whom to go to the movies with. Whose car to ride home in. Whether to watch the tube tonight, or read a book on investing. But, and you can be sure of this, each choice counts. Each choice is a building block - some large, some small. But each one is a part of the structure of your life. If you make the right choices, or if you make more right choices than wrong ones, something absolutely terrible may happen to you. Something unthinkable. You, my friend, could become one of the hated, the evil, the ugly, the feared, the filthy, the successful - the rich.

Quite a few people have made that mistake.

The rich basically serve two purposes in this country. First, they provide the investments, the investment capital, and the brains for the formation of new businesses. Businesses that hire people. Businesses that send millions of paychecks home each week to the un-rich.

Second, the rich are a wonderful object of ridicule, distrust, and hatred. Few things are more valuable to a politician than the envy most Americans feel for the evil rich. Envy is a powerful emotion. Even more powerful than the emotional minefield that surrounded Bill Clinton when he reviewed his last batch of White House interns. Politicians use envy to get votes and power. And they keep that power by promising the envious that the envied will be punished: 'The rich will pay their fair share of taxes if I have anything to do with it.' The truth is that the top ten percent of income earners in this country pays almost fifty percent of all income taxes collected. I shudder to think what these job producers would be paying if our tax system were any more "fair."

You have heard, no doubt, that in the rich get richer and the poor get poorer. Interestingly enough, our government's own numbers show that many of the poor actually get richer, and that quite a few of the rich actually get poorer. But for the rich who do actually get richer, and the poor who remain poor… there's an explanation - a reason. The rich, you see, keep doing the things that make them rich; while the poor keep doing the things that make them poor.

Speaking of the poor, during your adult life you are going to hear an endless string of politicians bemoaning the plight of the poor... so you need to know that under our government's definition of "poor" you can have a five million dollar net worth, a $300,000 home and a new $90,000 Mercedes, all completely paid for. You can also have a maid, cook, and valet, and one million in your checking account, and you can still be officially defined by our government as "living in poverty." Now there's something you haven't seen on the evening news!

How does the government pull this one off? Very simple, really. To determine whether or not some poor soul is "living in poverty," the government measures one thing - just one thing. Income. It doesn't matter one bit how much you have, how much you own, how many cars you drive or how big they are, whether or not your pool is heated, whether you winter in Aspen and spend the summers in the Bahamas, or how much is in your savings account. It only matters how much income you claim in that particular year. This means that if you take a one-year leave of absence from your high-paying job and decide to live off the money in your savings and checking accounts while you write the next great American novel, the government says you are "living in poverty."

This isn't exactly what you had in mind when you heard these gloomy statistics, is it? Do you need more convincing? Try this. The government's own statistics show that people who are said to be "living in poverty" spend more than $1.50 for each dollar of income they claim. Something is a bit fishy here. Just remember all this the next time Katie Couric tells you about some hideous new poverty statistics.

Why has the government concocted this phony poverty scam? Because the government needs an excuse to grow and to expand its social welfare programs, which translates into an expansion of its power. If the government can convince you, in all your compassion, that the number of "poor" is increasing, it will have all the excuse it needs to sway an electorate suffering from the advanced stages of Obsessive-Compulsive Compassion Disorder.

I'm about to be stoned by the faculty here. They've already changed their minds about that honorary degree I was going to get. That's okay, though. I still have my Ph.D. in Insensitivity from the Neal Boortz Institute for Insensitivity Training. I learned that, in short, sensitivity sucks. It's a trap. Think about it - the truth knows no sensitivity. Life can be insensitive. Wallow too much in sensitivity and you'll be unable to deal with life, or the truth. So, get over it.

Now, before the dean has me shackled and hauled off, I have a few random thoughts.

You need to register to vote, unless you are on welfare. If you are living off the efforts of others, please do us the favor of sitting down and shutting up until you are on your own again.

When you do vote, your votes for the House and the Senate are more important than your vote for president. The House controls the purse strings, so concentrate your awareness there.

Liars cannot be trusted, even when the liar is the President of the United States. If someone can't deal honestly with you, send them packing.

Don't bow to the temptation to use the government as an instrument of plunder. If it is wrong for you to take money from someone else who earned it - to take their money by force for your own needs - then it is certainly just as wrong for you to demand that the government step forward and do this dirty work for you.

Don't look in other people's pockets. You have no business there. What they earn is theirs. What you earn is yours. Keep it that way. Nobody owes you anything, except to respect your privacy and your rights, and to leave you the hell alone.

Speaking of earning, the revered Forty-hour workweek is for losers. Forty hours should be considered the minimum, not the maximum. You don't see highly successful people clocking out of the office every afternoon at five. The losers are the ones caught up in that afternoon rush hour. The winners drive home in the dark.

Free speech is meant to protect unpopular speech. Popular speech, by definition, needs no protection.

Now, if you have any idea at all what's good for you, you will get the hell out of here and never come back. Class dismissed."

A BAND OF GOLD

"The fact that nobody wants to believe something doesn't keep it from being true."

In the late 1980s my platoon was preparing for a winter operation in Norway called "Cold Winter," and as part of that preparation we deployed to the Marine Corps Mountain Warfare Training Center in Bridgeport, California, high up in the Sierra Nevada Mountains.

After spending the initial weeks honing our skiing and winter survival skills we moved into a tactical phase, during which we were tasked with actual reconnaissance missions in a winter environment. Anyone who has ever done that sort of thing knows how challenging it can be. One night our mission was to conduct a point recon of an enemy encampment which was to be assaulted by the infantry the following morning.

After being inserted by helicopter a few miles from the objective my team skied quietly through the snow, stopping only for security halts and map checks. It was a moonless night, so during those map checks I had to conceal myself under a poncho and use a red lens flashlight in order to maintain light discipline. I soon discovered it was virtually impossible to manipulate the map, light and poncho while wearing gloves, so I pulled them off to get a better feel for things - and I think that was the biggest problem I encountered the entire night. In the end the mission turned out to be quite successful, and we returned to the LZ for extraction and debrief.

Once we had returned to base the entire team gathered to be debriefed by the battalion commander, Lieutenant Colonel Wayne Rollings. As a former Force Recon CO, Colonel Rollings was one of the few officers who knew how to properly employ his reconnaissance assets, and we loved working for him. As the meeting began I removed my gloves for the first time since coming off the mountain, and immediately knew something was not right. Back then I had a habit of spinning my wedding ring on my finger, and I was surprised to find it wasn't there. It occurred to me that I'd experienced significant weight loss during the preceding weeks due to the combination of physical exertion and cold weather, and had noticed that everything was fitting much more loosely - including the ring. I frantically checked inside my left glove, but found nothing. Then I thought back to the night on the mountain. I had been wearing the ring when we boarded the chopper, so the only explanation was it had come off during one of the map checks. I wasn't pleased, but what could I do.

At the end of our training in Bridgeport we were taken to Reno, Nevada for a couple of days of much deserved liberty before moving on to Fort McCoy, Wisconsin (and later Norway) for a few more weeks. Reno, known as the "Biggest Little City in the World," is of course filled with casinos much like Las Vegas. At that time it was also home to the world famous, but now defunct, Mustang Ranch. I am not a gambler and quickly grew bored hanging out at the blackjack table watching other people play, so when someone suggested we visit that famous landmark I figured why not - I could just sit at the bar and have a drink while the others "sampled the wares" in the back.

When we got there and it came time to pay the taxi driver one of the others, Sergeant 'Brick,' discovered he had no cash, but he assured me that if I paid his share he would buy me a drink on his credit card when we got inside. That's what we did, except when the bar bill came his card was rejected for being over-limit. That's when Brick did one of the stupidest things I have ever witnessed. He got on the phone, called his wife back in North Carolina, and told her he was trying to use his credit card at the Mustang Ranch! Then, to add insult to injury, he told her I was with him - and since our wives were close friends it wasn't long before I was in the doghouse. Naturally, she had assumed the worst.

When we returned home a few weeks later I had to explain to my then-wife that I had somehow managed to lose my wedding ring, and since she was still stewing about the Mustang Ranch episode it was just like throwing gasoline on a fire. But perhaps that's a bad analogy, since the look she gave me was colder than the glaciers I had been living on.

WITHOUT
Fault, Merit or Need

"I'd give a million dollars to be a Marine."
– Former heavyweight champion Riddick Bowe

Readers of "Swift, Silent and Surrounded" will recall a piece entitled "Watch Your Wallet," which dealt with the injustices of the Uniformed Services Former Spouses Protection Act (USFSPA) - the law that divvies up a retired servicemember's retirement pay and awards a significant portion to his or her former spouse. While many readers were horrified to learn of this law, a few commented that it was probably necessary in order to provide for spouses who had loyally followed the military way of life for a lot of years. While I totally agree that there are cases where that is true, it should be for a judge to decide on a case by case basis. Under the current law, courts feel compelled to award a division of this "property" almost automatically regardless of fault, merit, circumstance or financial need, and this practice needs to stop for the reasons outlined in the aforementioned story. But since I can probably talk about this in abstract terms until I am blue in the face, let me offer a few real life examples from around the country so you can draw your own conclusions:

From California: An Air Force Colonel was taken prisoner-of-war, sent to a North Vietnamese POW camp in the Fall of 1967, and was repatriated to the U.S. Government in 1973. Shortly after returning home after six years in captivity, he was served with divorce papers. According to those papers, the court declared the "date of separation" from his spouse to be April 1, 1970 - while he was still a prisoner-of-war! The former spouse did not have to repay any pay and allowances she received and spent AFTER the "date of separation"; she was also entitled to his accrued leave pay, and to monies he received under the War Crimes Act for inhumane treatment. The former spouse was also awarded the home, car, 42.7% of the member's military retired pay, child support and spousal support (even though the former spouse had numerous open affairs during the Colonel's incarceration in a POW camp and ended up marrying the attorney who prepared the divorce action on her behalf).

From California: A Marine Corps Staff Sergeant, returning to his duty station in Twenty-Nine Palms after serving in combat during Operation Desert Storm, planned to retire with twenty years of honorable military service. Upon his arrival home, his wife of nineteen years was found cohabitating with another man. In May of 1991 the spouse abandoned the service member and their three children and filed for no-fault divorce in California. The divorce was final in January of 1992. The military member was ordered to give 50% of the property of the marriage to the former spouse, and 47.5% of his military retired pay.

From Alaska: An Air Force Master Sergeant served twenty years in the military, including two tours in Vietnam. He and his wife were married the final sixteen years of his military service. While stationed in Alaska, and entering his last year before retirement, he was sued for divorce by his wife (who had found a boyfriend) and evicted from his home. The court awarded the ex-spouse 40% of the service member's retired pay as property, and an additional 27% as child support. After taxes, the retired service member receives approximately $130 monthly. Keep in mind the former spouse was employed at $34,000 per year; and her live-in boyfriend was employed at $26,000 per year; the military retiree pays her $9,000 per year.

From California: A Navy Chief Petty Officer (E-7), who was a Vietnam Veteran, was divorced in 1978 after eleven years of marriage. At the time of divorce, real property was awarded along with child support. Thirteen years later, in 1991, the military member retired as a Master Chief Petty Officer with thirty years of service. At that time, the former spouse returned to court for division of retired pay and was awarded 28% payable at the E-9 rate. At the time of this award, the former spouse was married to her fourth husband.

From Virginia: During a ten year marriage, an Army Captain financed a college education and master's degree for his unemployed spouse so she could obtain a career position. The former spouse then filed for divorce from the military member. Ten years after the divorce, the service member must pay a portion of his retired pay to the spouse as awarded by the court. There were no children, and both tangible property and savings were given to the former spouse (who is gainfully employed, remarried and highly financially secure).

From Louisiana: A female civilian is presently in her fourth marriage to a retired military member. She is collecting USFSPA payments from the first three ex-spouses.

From Wisconsin: A female member of the Navy married a male enlisted shipmate who brought two children from an earlier marriage into the relationship. Shortly thereafter, her husband was discharged from the Navy for drug abuse. Subsequently, they had a child of their own. The wife was the principal family provider throughout most of the marriage. They were divorced in 1994, and when she retired as a First Class Petty Officer with twenty years of service her husband, who had remarried and had been in and out of jail, was awarded 30% of her retired pay for life.

If you can see nothing wrong with those scenarios, I hope it happens to you!

THE LAW OF WAR

"A good name, like good will, is attained by many actions and may be lost by only one."

One of the most important parts of Marine Corps Essential Subjects (EST) training teaches the Law of War, which dictates how we are supposed to conduct ourselves on the battlefield. They say all is fair in love and war, but that isn't exactly true. Combat, while brutal, does have guidelines of gentlemanly conduct, and any nation that fails to observe them is considered a terrorist state. I recall the EST training I received as being rather dry and boring - and suggest the words in the following speech be used as an example of how to conduct oneself.

With one phrase, Lieutenant Colonel Tim Collins, commander of the 1st Battalion of the Royal Irish, summed up the task in hand for the British forces waiting to remove Saddam Hussein from Iraq in March 2003.

Collins, known as "Nails" by his men for his tough-guy attitude, was addressing his eight hundred men, an arm of Britain's 16 Air Assault Brigade, at Fort Blair Mayne, a Kuwaiti desert camp twenty miles south of the Iraqi border.

Two months later, a month after the dramatic liberation of Iraq, Collins - now promoted to full Colonel - hit the headlines again. He was accused by a junior U.S. officer of brutality towards Iraqi civilians, the very crime he warned against in his famous speech. He had allegedly shot at the tires of vehicles when not under threat, and pistol-whipped and shot at the feet of civic leaders.

Collins strenuously denied the charges. He admitted to shooting out the tires of looters' vehicles to stop them from making off with essential supplies, and shooting into the kitchen floor of a senior Ba'ath party member to jog his memory about where he had hidden his weaponry. All part of a robust approach, Collins said, to imposing psychological as well as physical dominance of the enemy. He categorically denied charges of pistol-whipping or beating prisoners. It appeared Collins and his accuser had a personality clash while working together, leading to the accusation.

Judge for yourself the character of this man. Here is the full text of his extraordinary and electrifying speech:

"We go to liberate, not to conquer. We will not fly our flags in their country. We are entering Iraq to free a people, and the only flag which will be flown in that ancient land is their own. Show respect for them.

There are some who are alive at this moment who will not be alive shortly. Those who do not wish to go on that journey, we will not send. As for the others I expect you to rock their world. Wipe them out if that is what they choose. But if you are ferocious in battle, remember to be magnanimous in victory.

Iraq is steeped in history. It is the site of the Garden of Eden, of the Great Flood, and the birthplace of Abraham. Tread lightly there. You will see things that no man could pay to see, and you will have to go a long way to find a more decent, generous and upright people than the Iraqis. You will be embarrassed by their hospitality even though they have nothing. Don't treat them as refugees for they are in their own country. Their children will be poor, and in years to come they will know that the light of liberation in their lives was brought by you.

If there are casualties of war then remember that when they woke up and got dressed in the morning they did not plan to die this day. Allow them dignity in death. Bury them properly and mark their graves.

It is my foremost intention to bring every one of you out alive, but there may be people among us who will not see the end of this campaign. We will put them in their sleeping bags and send them back. There will be no time for sorrow.

The enemy should be in no doubt that we are his nemesis and that we are bringing about his rightful destruction. There are many regional commanders who have stains on their souls and they are stoking the fires of hell for Saddam. He and his forces will be destroyed by this coalition for what they have done. As they die they will know their deeds have brought them to this place. Show them no pity.

It is a big step to take another human life. It is not to be done lightly. I know of men who have taken life needlessly in other conflicts, and I can assure you they live with the mark of Cain upon them. If someone surrenders to you then remember they have that right in international law and ensure that one day they go home to their family.

The ones who wish to fight, well, we aim to please.

If you harm the regiment or its history by over-enthusiasm in killing or in cowardice, know it is your family who will suffer. You will be shunned unless your conduct is of the highest, for your deeds will follow you down through history. We will bring shame on neither our uniform nor our nation.

As for ourselves, let's bring everyone home and leave Iraq a better place for us having been there.

Our business now is north."

The preceding speech was given by Lieutenant Colonel Tim Collins, 20 March 2003, in Kuwait near the Iraqi border.

IT'S NOT ABOUT SEX

Eric Jowers

"Nearly all men can stand adversity, but if you want to test a man's character, give him power." - Abraham Lincoln

I often find myself getting into conversations during which people defend the actions of former President Clinton by saying what he did in his personal life was his own business and had nothing to do with his ability to function as a leader. Anyone who believes that does not know the first thing about leadership, and certainly has never been subject to the Uniform Code of Military Justice. If a General were to engage in similar inappropriate conduct with a Lance Corporal, I wonder if the Pentagon would say his personal life is "his business?" Of course not. That is why we in the military are not inclined to look the other way.

There has been much media coverage concerning Article 88 of the Uniform Code of Military Justice and how it might apply to an active duty Marine ever since Major Shane Sellers published a very controversial article about then-President Clinton in the Navy Times. The question is, did Major Sellers have the right to exercise his First Amendment rights under the United States Constitution? And, should the Commander-in-Chief of the Armed Forces be held to the same standards as his military commanders?

A letter on the subject was written to then-President Clinton by Eric Jowers, a retired Army officer, which I believe says it all:

Dear Mr. President,

It's not about sex. If it were about sex, you would be long gone. Just like a doctor, attorney or teacher who had sex with a patient, client or student half his age, you would have violated the ethics of your office and would be long gone. Just like Sergeant Major of the Army Gene McKinney who, though found not guilty, was forced to resign amid accusations of sexual abuse. Remember the Air Force General you wouldn't nominate to be Chairman of the Joint Chiefs of Staff because he freely admitted to an affair almost fifteen years before, while he and his wife were separated? Unlike you, he was never accused of having a starry-eyed office assistant my daughter's age perform oral sex on him while he was on the phone and his wife and daughter were upstairs.

If it were about sex, you should be subjected to the same horrible hearings that Clarence Thomas was subjected to because of the accusations of Anita Hill. The only accusation then was that he talked dirty to her. He didn't even leave semen stains on her dress.

No, it's not about sex. It's about character. It's about lying. It's about arrogance. It's about abuse of power. It's about dodging the draft and lying about it. When caught in a lie by letters you wrote, you concocted a story that nobody believed. But we excused it and looked away.

It's about smoking dope, and lying about It. "I didn't inhale," you said. Sure, and when I was sixteen and my buddies and I swiped a beer from an unwatched refrigerator, we drank from it, but we didn't swallow. "I broke no laws of the United States," you said. That's right, you smoked dope in England or Norway or Moscow, where you were demonstrating against the U.S.A. You lied, but we excused it and looked away.

It's about you selling overnight stays in the White House to any foreigner or other contributor with untraceable cash.

It's about Whitewater and Jim and Susan McDougal and Arkansas Governor Jim Guy Tucker and Vincent Foster and Gennifer Flowers and Paula Jones and Kathleen Willey and nearly countless others.

It's about stealing the records from Foster's office while his body was still warm and putting them in your bedroom and "not noticing them" for two years.

It's about illegal political contributions. It's about you and Al Gore soliciting contributions and selling influence at Buddhist temples and in the same Oval Office where Abraham Lincoln and Franklin Roosevelt led their countries through the dark days of wars that threatened the very existence of our nation. But we excused you and looked away.

It's about hiding evidence from Ken Starr, refusing to testify, filing legal motions, coaching witnesses, obstructing justice and delaying Judge Starr's inquiry for months and years, and then complaining that it has gone on too long. The polls agreed. Thank goodness that Judge Starr didn't read the polls, play politics or excuse you and look away. He held on to the evidence like a tenacious bulldog.

Your supporters say that you've confessed your wrong doings and asked for our forgiveness. Listen, what you said on TV the night you testified to the grand jury was not a confession. Confession in the face of overwhelming evidence is not a confession at all. Not that it would make a lot of difference. A murderer who contritely confesses his crime is still a murderer.

When your "confession" didn't sell, even to your friends, you became more forthcoming. Maybe someday you'll confess more, but probably not. You've established such a pattern of lying that we can't believe you anymore. Neither can your cabinet, the Congress or any of the leaders of the nations of the world. When a leader's actions defame and emasculate our country as profoundly as yours have, it's no longer a personal matter, as you claim. It's no longer a matter among you, your family and your God.

By the way, I don't believe for a minute that Hillary was unaware of your sexual misadventures, abuse of power and pattern of lying. She has been a party to your wrongdoings since Whitewater and Gennifer Flowers, just as surely as she lied about the Rose law firm's billings and hid the Vincent Foster evidence in your bedroom for two years. Why? So she could share in the raw power that your office carries. The two of you probably lied to Chelsea, but that is a matter among you, your family and your God.

Remember the sign over James Carville's desk during the 1992 campaign? It said, "It's the economy, stupid!" Place this sign over *your* desk: "It's about character, stupid!"

No, it's not about sex, Mr. President. If it were, you would be long gone. It's about character, but we had to live with your filth, lies and arrogance for a while longer. Your lies, amorality and lack of character were as pervasive as they were despicable, so we had no reason to believe you would quietly resign and go away.

You counted on half truths and spin doctors to see you through, the country be damned. It always worked before. We excused you and looked the other way. You made every elected official, minister, teacher, diplomat, parent and grandparent in the country apologize for you and explain away your actions. And when you left, your legacy to the United States of America was be a stain on the Office of the President that was as filthy as the stain on Monica's dress. It took a lot of scrubbing to make it clean again."

SERGEANT MAXEY

By Captain Jason D. Grose

"Example IS Leadership." - Dr. Albert Schweitzer

Over the last four years I have been exposed to near toxic levels of leadership training. I have read papers, written papers, listened to speeches, and had numerous discussions on just about every aspect of the subject. I have learned a great deal about textbook leadership but in order to spare you, the reader, from another idealistic, theoretical, dry overview of the subject, I want to provide a different perspective. While I do not claim to have all of the answers, my quest for the ideal has taught me many lessons I feel qualified to share. Because the first tenant of leadership is to lead by example, the subject in this paper describes the greatest influence to my concept of leadership. Through his examples, you can see the roots of my leadership beliefs.

If you were to ask any great leader who they idolized, he would probably be able to pick out one person who stands out among the hordes of contributing leaders he has encountered. Everyone has their heroes, and our interaction with these guiding points of light have profound effects on who we are.

Sergeant Shane Maxey is my mentor and the best leader I have ever known. To me, he is the very definition of leadership. Although tough as nails, Sergeant Maxey had a compassion for his people which, no matter how hard he tried to conceal it, shone through his rough exterior. But to understand this walking textbook on leadership, a little background information is necessary.

Sergeant Maxey was a hard cookie from the start. Due to a rather troublesome upbringing, he developed a hard-headed attitude which resulted in a critical and negative personality. After joining the Marine Corps, the institution's demand for excellence together with his own intense discipline created a chemical reaction. Suddenly it was as though the Corps was made for him, rather than vice versa. He excelled as a junior Marine, attaining rank and responsibility quickly. Maxey eventually became a drill instructor, and spent three successful years as arguably the best DI in San Diego. After completing his time on the drill field he returned to avionic maintenance, where he had served prior to volunteering for DI duty. At this point our paths crossed, and I would never again be the same.

It is said that the best kind of leadership is not out of any book. Leadership by example is the only real way to learn. In this spirit, a few pivotal examples come to mind which show why Sergeant Maxey holds such an important place in my leadership model.

The first time I met Sergeant Maxey, I hated him. Avionics had always been a pretty lax environment with few harsh conflicts. I had "grown up" with this attitude and was professionally deficient. The day Sergeant Maxey checked in, he still mentally had his DI cover on. There was a call for a FOD (Foreign Objects and Debris) walk on the avionics pad and I was hiding in the back of the shop, sipping on some coffee. I thought I could get away with hiding out if I looked busy. Maxey walked by, stopped, and pointed a finger at me. "WHAT GOOD ARE YOU DOING?" My deer-caught-in-the-headlights response was accompanied by some unconvincing mumbling, prompting Sergeant Maxey to bark, "GET YOUR BUTT OUT THERE WITH THE REST OF THEM!" Thus began our relationship.

Looking back, the lesson I learned from that experience was that as a leader, do not hesitate to lead even if you are new. Sergeant Maxey took charge from the moment he arrived. None of us knew it, but he was the number one sergeant in the avionic division before he went the drill field, and thus was also very technically proficient. From the beginning he expected the most out of those who were put in his charge, asking from them what he gave: 100%.

For the next few months Sergeant Maxey continued to be the disciplinary force for our workshop. He made no friends, but strange things started to happen - the quality of our work increased, and the overall expectations around the work center increased. It was fate that made this happen because the Gulf War was just beyond the horizon.

When the work center Staff NCOs called Sergeant Maxey into their office, not one of them could look him in the eyes. He even intimidated most of his seniors, and what they had to tell him they knew he would not like. He knew exactly what was going on but no one wanted to tell him he would be leading the upcoming deployment. He had served his time on the drill field and had just reunited with his wife after arriving early to Yuma in order to set things up. He had given a lot to the Marine Corps in the past three years and felt he needed a little "down-time." But this would not be the case. Knowing it was futile to fight the assignment and knowing he was best qualified for the position, he told the staff he would lead it only if he could pick the Marines that went with him. Such was his power and brashness. This demand, unheard of from a sergeant to a group of Staff NCOs, was granted - and I was among those he picked.

This is the next lesson I learned from Sergeant Maxey. As a leader, you owe it to your people to be forceful yet respectful to superiors. As rough as he was, I never once heard him fail to start a sentence with "Sir" when speaking to an officer. But he never backed down, even in the face of conflict with higher echelons. He was a force to be reckoned with, but somehow kept an air of respect even with those who did not deserve it. I found this combination of forceful attitude and professional respect astounding.

The night before we deployed, Sergeant Maxey invited those of us who had wives over to his house. Even though it was his anniversary, he invited all of us to sip wine and spend the evening together in the spirit of camaraderie. That night he made a promise to the wives, unknown to the husbands, that he would bring us all back alive. This promise he took very seriously, and eventually kept.

Later that night I asked if I could speak with him. I was having trouble figuring out what I was going to tell my mother. The deployment was pretty obvious to the public, but we were not allowed to call loved ones because of security considerations. He took me to his phone in a back room and not only told me to call her, but also what to tell her. He told me to tell her I had to go away on business for awhile, and that she knows what business I was in. I was to tell her I would write when I could and that I would be home soon. He then left the room. I called my mother and followed his instructions. She cried, and I then called my brother, an ex-Army soldier, and told him the same story. He understood, told me to get home in one piece, and said he would call Mom to calm her down.

Sergeant Maxey taught me two things that night. First, the line between a leader's personal life and professional life is transparent. Second, a leader's responsibility extends to the families of those led. Sergeant Maxey invited us into his home that night despite the fact it was his wedding anniversary. A lesser man would have justifiably spent the night alone with his spouse, especially considering the impending deployment. But not him. He extended his family boundaries to those under his charge, and that had a profound effect on me. By making commitments to the wives, he in essence took personal responsibility for the Marines' safety. This was not required, and is not written in any book. Yet this example of leadership perfectly demonstrates everything ever written on the subject.

One of the best leadership lessons I absorbed was how much one individual can affect a group. Sergeant Maxey is the kind of person people sit around and tell legendary stories about - when two people get together that knew him, much time is spent telling stories of his greatness.

As a leader, Sergeant Maxey did an amazing thing. He redefined what was "cool" in our workshop. Before he came many of the young, single Marines bragged of drinking, womanizing, and general mischief. The more women they dated, the higher their status was within the group. Sergeant Maxey never publicly admonished the practice, but his influence changed that attitude over time.

Maxey had been married to the same woman for over ten years and they had two children. One day he invited the shop over to his house for a barbecue, and I arrived early to help. What not many people knew was Sergeant Maxey did exactly half of the housework at home. He would say, "I cause at least half the mess, probably more, so why shouldn't I help clean it up?" Every Saturday he, his wife Michelle, and the kids would spend the better part of the day cleaning the house. When the group showed up for the barbecue, Sergeant Maxey was just finishing up and was cleaning the downstairs toilet. They all walked in and saw him hunched over, scrubbing away. Needless to say there were shocked gasps, snickering, and outright laughter. There, on his hands and knees, was the roughest, hardest, most fire-breathing sergeant in the United States Marine Corps, scrubbing away on a toilet like a private. With all the dignity in the world, he stood up, dropped the scrubber, and silently dared anyone to say anything. It was a moment I will never forget. I know he redefined many attitudes that day.

The lesson was that no one, not even the most senior leader, is above menial work. He showed that a real man is not the one with the most notches on his ego, but a hard-working, faithful, fair leader who is willing walk the walk both at work and at home.

His sea stories also include how he won a bet with the men in his shop on his first deployment. When they bet he could not stay faithful in the Philippines, Sergeant Maxey defined appropriate behavior without issuing a single order. The high degree of respect he earned from the men of the shop helped to redefined beliefs. Suddenly it was the "in" thing to be monogamous because that was how Sergeant Maxey operated. By no means was there a blind hero-worship attitude in the shop because too many independent personalities were involved for that. We did not gush over Sergeant Maxey, and even hated him sometimes. But whether you liked him or hated him, you were affected by him. The definition of being a "real man" became a monogamous, family-oriented, professional, moral, ethical, and proficient Marine.

One of the final effects of Sergeant Maxey's leadership resulted in my commission. From the beginning of my career, I had planned on becoming an officer. Ignorantly, I laid out a plan from the start to enlist and then just pick up a commissioning program. Big talk from a nineteen-year old PFC. I did not consider the work it would take, and stupidly bobbed along like the Marine Corps owed me a slot in

MECEP. After the first two years, my dream was fading. I had always been an average Marine, but never excelled. Liberty was more important than my job, and my undemanding role as an avionics technician did not challenge my abilities. I became comfortable… too comfortable.

After I met Sergeant Maxey I realized how deficient I was becoming. I started to feel uncomfortable with my surroundings, and as my expectations increased I started to feel the need to improve myself. As I made these realizations, my attitude resulted in alienation from my contemporaries. My need to exercise leadership clashed head-on with the status quo of those around me. I was confused about what to do and started believing I would make a lousy officer.

I had never told Sergeant Maxey my intentions to become an officer, and by the time the Gulf War rolled around I had almost forgotten my desire for a commission. But after the deployment Sergeant Maxey told me I should apply, and after I told him about my original goal there was no stopping his involvement.

By that time I had lost confidence in my leadership abilities. I had managed to alienate myself from just about everyone around me, but Sergeant Maxey saw something different - and I often tell people there is a permanent boot mark on my butt the size of Sergeant Maxey's foot. He hounded me, threatened me, yelled at me, and pushed me to put together my application. He believed in me when I did not believe in myself. With his prodding I realized I could be a great officer, and I owed it to the Marines I would lead in the future to become one.

While sitting in a van in the communications center parking lot in Okinawa, I was told that I had earned a slot for the MECEP. Stunned, I sat motionless as the realization set it. Everyone in my shop, including myself, was surprised I had made it on the first try. Everyone, except one person. To Sergeant Shane Maxey, it was a foregone conclusion.

I was very lucky my path crossed his. Sergeant Maxey played a pivotal role in my life as a person, and as a Marine. Through professional and personal example, this exceptional leader demonstrated a style of leadership I now believe in. Whenever I am met with a difficult decision, the last check I make is to ask myself how Sergeant Maxey would handle it - and if my decision meets that standard, I know it is the right one. My single professional goal is to be such an example to someone in my life. No man could ask for more, and only then will I consider myself a true leader.

PRESIDENTIAL TRADITION

"A great leader never sets himself above his followers, except in carrying responsibilities."

Whenever TV news showed Marine One touching down on the White House lawn during the Clinton years I always found it quite ironic to watch the President return the salute of the Marines flanking the helicopter door. I commented more than once that I was glad I was not one of them, because the idea of saluting a draft dodger is not something I could easily stomach. Those thoughts resurfaced recently when I came across an urban legend which claimed those Marines had purposely not rendered proper military courtesies during that period, but some research on the subject has shown that to be false. Marines may find a task distasteful, but they are professionals who understand the importance of properly carrying out their duties. That said, I have had occasion to speak with a number of Marines who served aboard Marine One, and to a man they said while they did in fact do their jobs in the prescribed manner, they hated doing it. There is an old adage which says "They will salute the rank, but they have to respect the man," and it was never truer than in this case. I do take some solace in knowing the tradition of saluting the crew was initiated by someone much more worthy of the honor - President Ronald Reagan:

President Reagan wrote, "I never ceased to enjoy reviewing our men and women in uniform and hope I started a new tradition for Presidents. As Commander in Chief, I discovered it was customary for our uniformed men and women to salute whenever they saw me. When I'd walk down the steps of a helicopter, for example, there was always a Marine waiting there to salute me. I was told Presidents weren't supposed to return salutes, so I didn't, but this made me feel a little uncomfortable. Normally a person offering a salute waits until it is returned, then brings down his hand. Sometimes, I realized, the soldier, sailor, Marine, or airman giving me a salute wasn't sure when he was supposed to lower his hand. Initially I nodded and smiled and said hello and thought maybe that would bring down the hand, but usually it didn't. Finally, one night when Nancy and I were attending a concert at the Marine Corps headquarters, I told the Commandant of Marines, "I know it's customary for the President to receive these salutes, but I was once an officer and realize that you're not supposed to salute when you're in civilian clothes. I think there ought to be a regulation that the President could return a salute inasmuch as he is Commander in Chief and civilian clothes are his uniform."

"Well, if you did return a salute," the General said, "I don't think anyone would say anything to you about it."

The next time I got a salute, I saluted back. A big grin came over the Marine's face and down came his hand. From then on, I always returned salutes. When George Bush followed me into the White House, I encouraged him to keep up the tradition."

Once the Clinton's had left the White House the act of saluting the President regained its meaning, and you could actually see the respect and admiration in the faces of those Marines once again. And it didn't take long for President George W. Bush to demonstrate why. On the television show

"Special Report" with Brit Hume they generally show a funny video clip at the close of the show, and while that clip is usually only a humorous diversion, there was one occasion where it carried a bit more meaning.

On that occasion they showed President and Mrs. Bush about to board Marine One on their way to Camp David for the weekend. As the video starts, the First Lady is leading the way into the helicopter with their Spaniel on a leash, and the President is right behind her with their Scotty. As Mrs. Bush entered the chopper the Marine posted at the crew chief's door saluted, and held his salute as the President approached. Right about then the Scottie the President was walking decided it wanted to sit, right when it got to the steps. The President pulled on its leash, but the stubborn dog persisted in sitting, so Mr. Bush bent down and scooped up the pooch and entered the waiting aircraft. Once the President had disappeared from view the Marine cut his salute and executed a facing movement in the prescribed manner. Moments later President Bush reemerged from the helicopter and out onto the steps. He saw that the Marine was standing at attention, head and eyes to the front, so the President leaned over and tapped him on the left arm. The startled Marine turned his body toward the President and received his return salute!

This simple act of respect for our military shows that President Bush really did get it. Most any other person of his stature, most notably the previous occupant of the White House, would have simply continued on their way and not bothered to return that salute. With that simple act he earned the respect of the military community.

CAPTAIN STROMBERG

By Capt Jason Grose

"It's not the load that breaks you down, it's the way you carry it." – Lena Horne

Why did I request Captain Stromberg to administer my oath? Who is this guy? Well, in honor of this most-impressive Marine, I would like to tell the story, as I remember it, of one of the bravest men I have ever met.

One day as we were preparing for the coming year our Marine Officer Instructor (MOI) at the time gathered us together in a formation and told us that we would soon have a visiting scholar. He said the captain was kind of funny looking, and for us not to be alarmed because he had been diagnosed with terminal brain cancer and would probably not be with us for the entire year. He was being assigned to the NROTC so that he could be near his family while he went through his medical treatments. Needless to say we were a little taken aback, and we expected the worst with news of a "walking dead man" joining our tight little group of future officers.

The next surprise was that his name was Captain Stromberg. I remembered that name from my days back in Yuma as an avionics technician. It turned out that his father, Major Stromberg, commanded one of the squadrons that MALS-13, my old logistical squadron, had supported.

When Captain Stromberg came on deck that summer he was nothing like what we had expected. Yes, the chemotherapy had taken its toll on the good Captain's hair, but his energy was astounding. Anything but a remorseful, dreading vision of gloom and doom, Captain Stromberg showed up with a more positive attitude and zest for life than anyone I had met at the University up until then. He did not want to be treated like a pitiful shell of a man. He was a Marine, and he was to be given all the respect that comes with being one. It would not take long before he would earn even more respect as a result of his attitude and sense of humor.

Throughout the year, Captain Stromberg had his ups and downs. Sometimes he would disappear for a few days, and when we asked we would be told he was recovering from some chemotherapy. When he returned he would be a little sluggish, and sometimes seemed a little disoriented. But what never changed was his attitude. He was always ready with a loud and boisterous greeting, usually followed by a hearty handshake, and never failed to brighten your day. More than once we could hear him walking from his office to the head just singing his lungs to their limit... inevitably off key. In the wardroom we would just look at each other, smile, and shake our heads.

When I left the unit, Captain Stromberg had been there two years. When he had first arrived, the doctors had given him only months to live. This was indicative of who this man was. Never did he give up, never did he cry on anyone's shoulder, never did he let his cancer dampen his strong and heroic personality. Because of this I walked into the MOI's office about a week before my commissioning, and asked if he would be offended if I asked Captain Stromberg to administer my oath. He was happy to forgo his traditional right as the senior Marine in the unit, so I marched down to Captain Stromberg's office and requested a word with him. As I entered he got right to his feet (a sign of respect I have never forgotten, and have incorporated into my own professional courtesy habits) and dropped the work he was in the middle of. I asked him if he would read me the oath at my upcoming commissioning. You would have thought that I asked him to be my best man at my wedding. Just the look on his face told me instantly it would be impossible to determine if the honor would mean more to me than to him, or vice versa. He was so happy and honored (I think the exact word he used was "stoked") about the idea. But he told me he had never done it before, and hoped I would not be offended if he read it off the paper because he did not have the entire oath memorized.

Less than a week later I was accompanied by my wife, two kids, mother, and brother into the Commanding Officer's office to receive my oath. As I stood there, so nervous that my knees were knocking, I was fortunate enough to have in front of me a Captain of Marines dressed in full dress blues administering my oath of office. Captain Stromberg stood there proudly, and I repeated every word he said, giving my word as a Marine that I would support and defend the Constitution of the United States of America. Those words, those promises that I made, came from the lips of one of the greatest Marines I have ever met - and when they came, they came from his heart because to my surprise, there were no notes. Captain Stromberg had memorized the entire oath and delivered it without so much as a glitch. This he had done for me, and it is just one of the reasons I consider my oath one of the most special moments of my life. Thank you, Captain Stromberg.

FALLEN COMRADES

In "Swift, Silent and Surrounded" I made mention of the Fallen Comrades table in an article about the traditions associated with Mess Nights, and some readers later asked for more information about what it symbolized. In response to that I request I now offer the following explanation as it would be provided by the President of the Mess, and dedicate it to the memory of First Sergeant Ed Smith, the Marines of Charlie Company, and all of our other fallen warriors. I think it is a fitting way to close out this volume:

As you entered the dining room tonight, you may have noticed a single table set in a place of HONOR.

It is a table set for each member of our Armed Services. Allow me to explain. Military tradition is filled with pride, customs and symbolism. This table is our way of symbolizing the fact that members of our profession of arms are missing from our midst. They are oftentimes called POWs, MIAs and KIAs. You remember them as your buddies. We call them Fallen Comrades.

The table setting is small, symbolizing the frailty of a prisoner alone against his oppressors.

The table cloth is black, symbolizing the uncertainty of our comrades' fate.

The place setting is white, this symbolizes the purity of our comrades' intentions to respond to his country's call to arms.

The single rose you see displayed reminds us of the families and the loved ones of our Comrades-in-Arms. Those who kept the faith, waiting for their return.

The red ribbon tied so prominently on the vase is reminiscent of the red ribbon worn upon the lapel and breasts of thousands and their unyielding determination to demand a proper accounting of our missing.

The slice of lemon on the bread place is to remind us of their bitter fate.

There is salt upon the plate, symbolic of the families' tears as they weep.

The glass is inverted, since they cannot toast with us tonight.

The chairs are empty, because they cannot be with us tonight.

ABOUT THE AUTHOR

Andy Bufalo retired from the Marine Corps as a Master Sergeant in January of 2000 after more than twenty-five years service. A communicator by trade, he spent most of his career in Reconnaissance and Force Reconnaissance units but also spent time with Amtracs, Combat Engineers, a reserve infantry battalion, and commanded MSG Detachments in the Congo and Australia.

He shares the view of Major Gene Duncan, who once wrote "I'd rather be a Marine private than a civilian executive." Since he is neither, he has taken to writing about the Corps he loves. He currently resides in Tampa, Florida.

Semper Fi

CPSIA information can be obtained
at www.ICGtesting.com
Printed in the USA
BVHW01s0259281217
503693BV00001B/30/P